Hydraulic Components Volume A

Hydraulic Sealing Elements

Dr. Medhat Kamel Bahr Khalil, Ph.D, CFPHS, CFPAI.

Director of Professional Education and Research Development,

Applied Technology Center, Milwaukee School of Engineering,

Milwaukee, WI, USA.

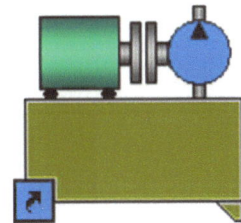

CompuDraulic LLC

www.CompuDraulic.com

CompuDraulic LLC

Hydraulic Components Volume A
Hydraulic Sealing Elements

ISBN: 978-0-9977634-9-2

Printed in the United States of America
First Published by August 2018
Revised September 2023

3850 Scenic Way, Franksville, WI, 53126 USA.

www.compudraulic.com

Disclaimer

It is always advisable to review the relevant standards and the recommendations from the system manufacturer. However, the content of this book provides guidelines based on the author's experience.

Any portion of information presented in this book might not be suitable for some applications due to various reasons. Since errors can occur in circuits, tables, and text, the author/publisher assumes no liability for the safe and/or satisfactory operation of any system designed based on the information in this book.

The author/publisher does not endorse or recommend any brand name product by including such brand name products in this book. Conversely the author/publisher does not disapprove any brand name product not included in this book. The publisher obtained data from catalogs, literatures, and material from hydraulic components and systems manufacturers based on their permissions. The author/publisher welcomes additional data from other sources for future editions. This disclaimer is applicable for the workbook (if available) for this textbook.

Table of Contents

PREFACE

This book provides a knowledge base for fluid power users to become familiar with commonly used seals in hydraulic components. This book presents an overview of hydraulic sealing elements including seal functions, classifications, and materials. This book also presents 15 various properties of hydraulic seals and relevant standard test methods. This book introduces best practices for hydraulic seals selection, installation and storage. This book provides a thorough analysis for failures of hydraulic seals including 26 failure modes with examples and demonstrative pictures. The following table shows some interesting statistics about the book.

Chapters	Pages	Lines	Words	Characters	Figures	Tables	Equations
11	167	1008	23795	133001	182	13	3

ACKNOWLEDGEMENT

The author likes to thank Mr. Thomas Wanke, CFPE Director of the Fluid Power Industrial Consortium at Milwaukee School of Engineering (MSOE), for his effective effort in reviewing the presented technical information.

The author also thanks the administration at Milwaukee School of Engineering and his supervisors who supported the effort to develop this book.

The author thanks the following companies (listed alphabetically) for permitting him to use portions of their copyrighted literatures in this book.

- American High-Performance Seals.
- ASSOFLUID
- Bar Hydraulics
- EPM Seals.
- Ecoseal Co. Ltd.
- Gates Corporation.
- Hallite Seals.
- Hydraulic Specialist Study Manual by IFPS
- MFP Seals.
- Parker Hannifin.
- System Seals Inc.
- Trelleborg.

Lastly, the author extends his thanks to the following sources of public information used to enrich the contents of the book.

- applerubber.com
- astonseals.com
- Caterpillar
- ecosealthailand.com
- hydrapakseals.com
- marcorubber.com
- mnrubber.com
- news.ewmfg.com
- o-ring-lab.com
- schoolcraftpublishing.com
- skf.com
- wyomingtestfixtures.com

Dr. Medhat Kamel Bahr Khalil

ABOUT THE AUTHOR

Medhat Khalil, Ph.D. is Director of Professional Education & Research Development at the Applied Technology Center, Milwaukee School of Engineering, Milwaukee, WI, USA. Medhat has consistently been working on his academic development through the years, starting from bachelor's and master's Degrees in Mechanical Engineering in Cairo Egypt and proceeding with his Ph.D. in Mechanical Engineering and Post-Doctoral Industrial Research Fellowship at Concordia University in Montreal, Quebec, Canada. He has been certified and is a member of many institutions such as: Certified Fluid Power Hydraulic Specialist (CFPHS) by the International Fluid Power Society (IFPS); Certified Fluid Power Accredited Instructor (CFPAI) by the International Fluid Power Society (IFPS); Member of Center for Compact and Efficient Fluid Power Engineering Research Center (CCEFP); Listed Fluid Power Consultant by the National Fluid Power Association (NFPA); and Listed Professional Instructor by the American Society of Mechanical Engineers (ASME). Medhat has balanced academic and industrial experience. Medhat has vast working experience in Fluid Power teaching courses for industry professionals. Being quite aware of the technological developments in the field of fluid power, Medhat had worked for several world-wide recognized industrial organizations such as Rexroth in Egypt and CAE in Canada. Medhat had designed several hydraulic systems and developed several analytical and educational software. Medhat also has considerable experience in modeling and simulation of dynamic systems using Matlab-Simulink. Medhat has been selected among the inductees for Pioneers in fluid Power by NFPA (2012) and Hall of Fame in fluid Power by IFPS (2021).

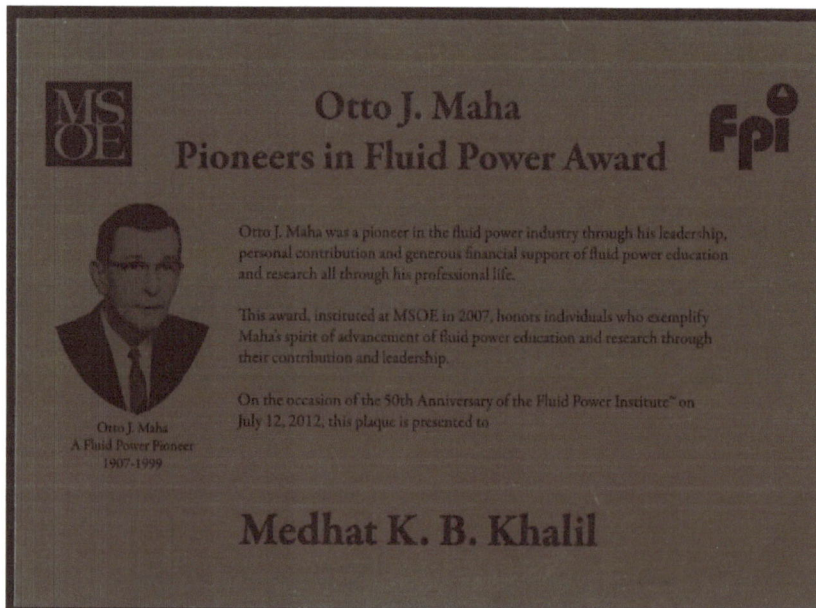

Chapter 1 – Introduction to Hydraulic Sealing Elements

The technology of producing hydraulic sealing elements and the types of seals are very broad. Additionally, sealing solutions vary widely depending on the hydraulic components, application, and working conditions. Another challenge is that manufacturers of hydraulic seals treat the production process as confidential to maintain their competitiveness in the marketplace.

1.1- Functions of Hydraulic Sealing Elements

Hydraulic seals are basically used to:

- Prevent external leakage and consequently:
  - Save money on oil. Assuming $10/Gallon is a current price, Figure A.1 shows significant money savings by having seals working properly.
  - Prevent environmental damage,
  - Improve work zone safety, and
  - Reduce accident liability due to fire or slip and falls.

- Prevent internal leakage from a high-pressure chamber to a low-pressure chamber and consequently:
  - Help components and systems function properly and efficiently.
  - Reduce heat generation.

- Provide controlled lubrication for adjacent parts or surfaces.

- Some hydraulic sealing elements have the primary function of preventing dirt, dust, and other contaminants from getting into the hydraulic components.

Leakage Rate	Monthly Losses	Yearly Losses
1 Drop/5Sec.	6.6 Gallon = $66	80 Gallon = $800
1 Drop/Sec.	34 Gallon = $340	409 Gallon = $4090
3 Drop/Sec.	113 Gallon = $1130	1243 Gallon = $12430
Steady Stream	720 Gallon = $7200	8640 Gallon = $86400

Spend tens of dollars to change the seal on time

Save hundreds of dollars on Troubleshooting

Save thousands of dollars on machine shutdown

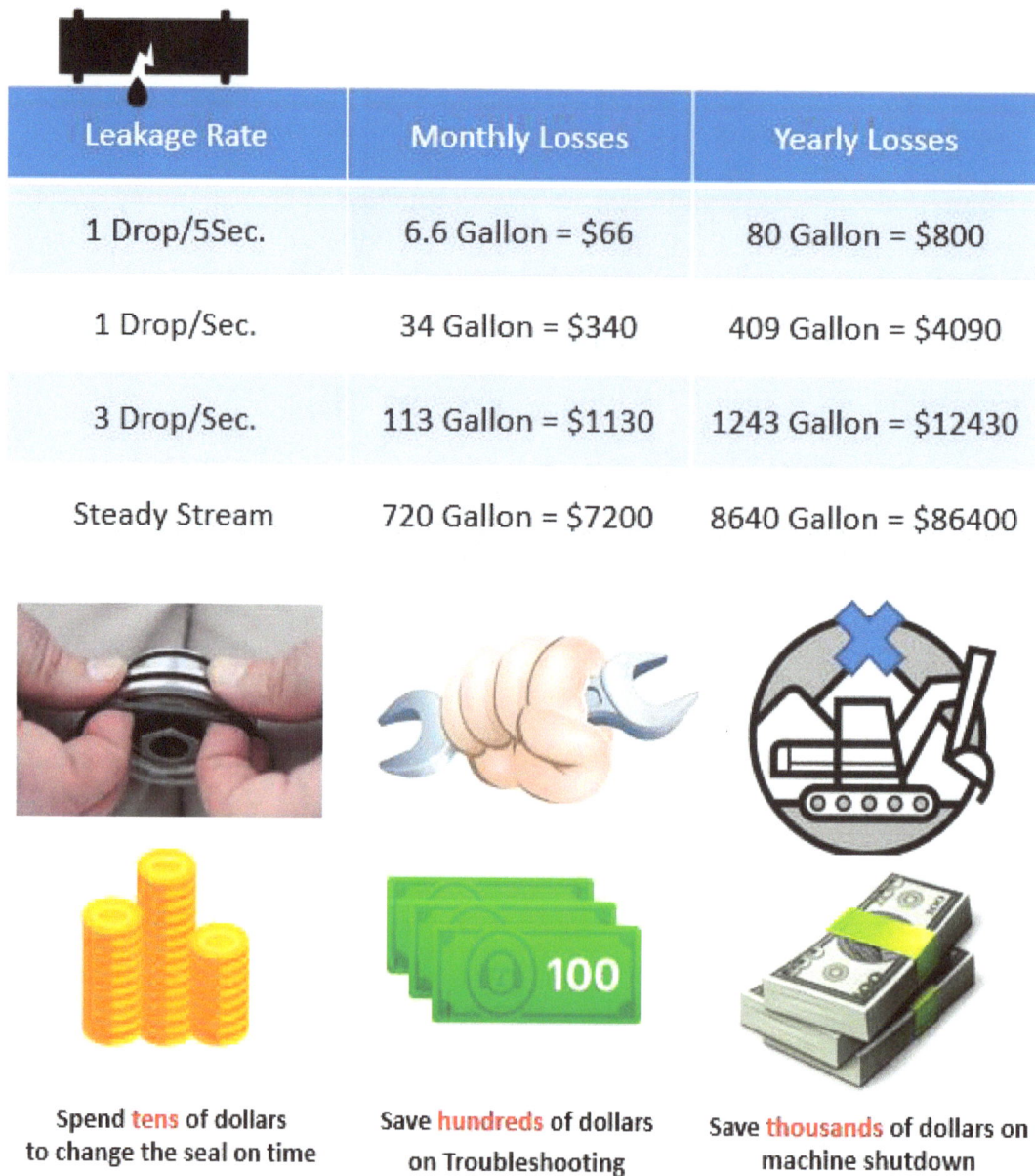

Fig. A.1 - Cost of Oil Leakage

1.2- Applications of Hydraulic Sealing Elements

On the systems level, hydraulic sealing elements are used in almost every application whenever a hydraulic system is used. Figure A.2 shows application examples where hydraulic seals are used. On the component level, hydraulic sealing elements are used in all components including pumps, actuators, valves, accumulators, etc.

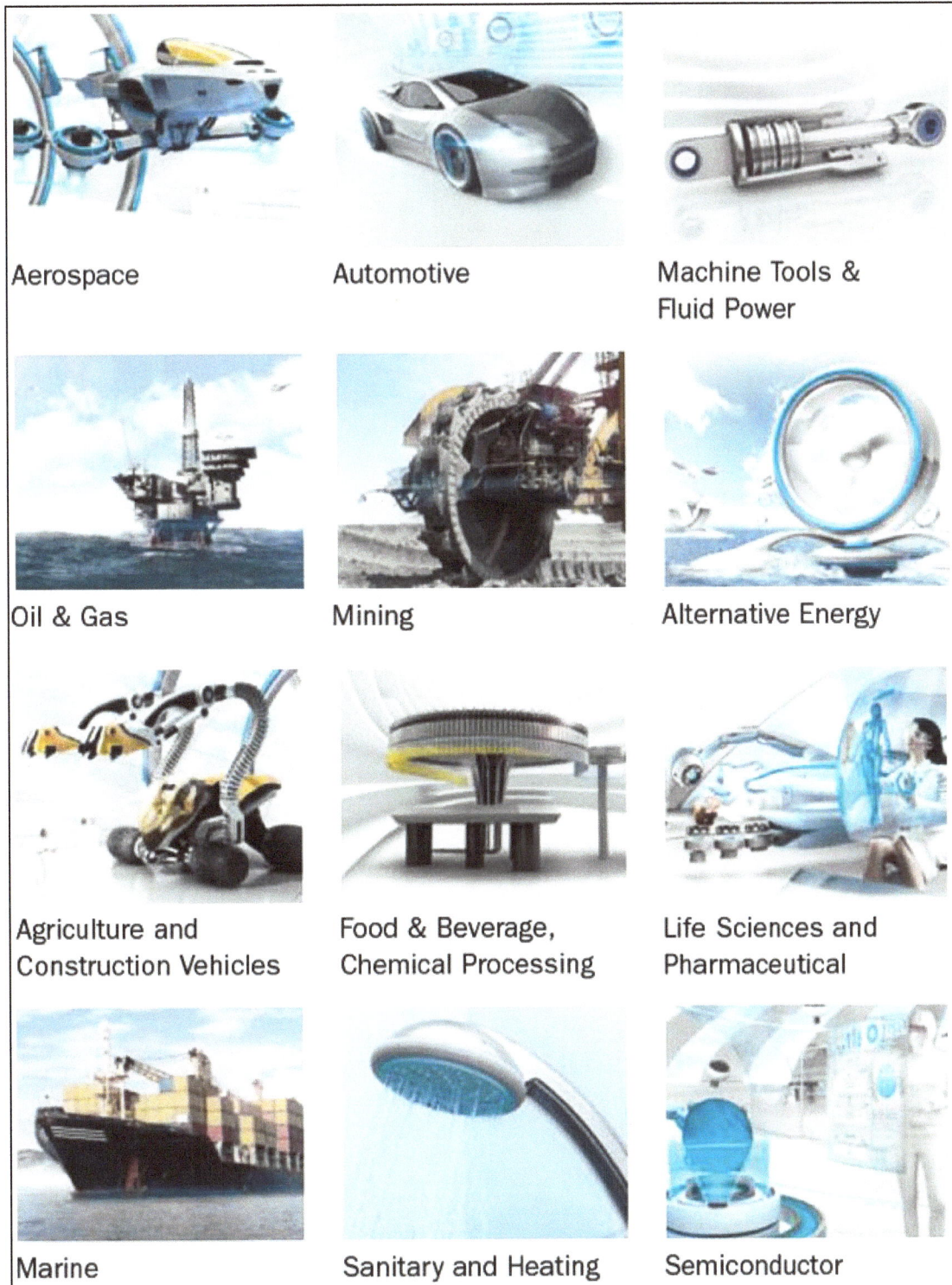

Fig. A.2 - Examples of Sealing Applications (Courtesy of Trelleborg)

1.3- Classifications of Hydraulic Sealing Elements

Figure A.3 shows the basic classifications of hydraulic sealing elements. Hydraulic sealing elements are broadly classified as *Static Seals* or *Dynamic Seals*.

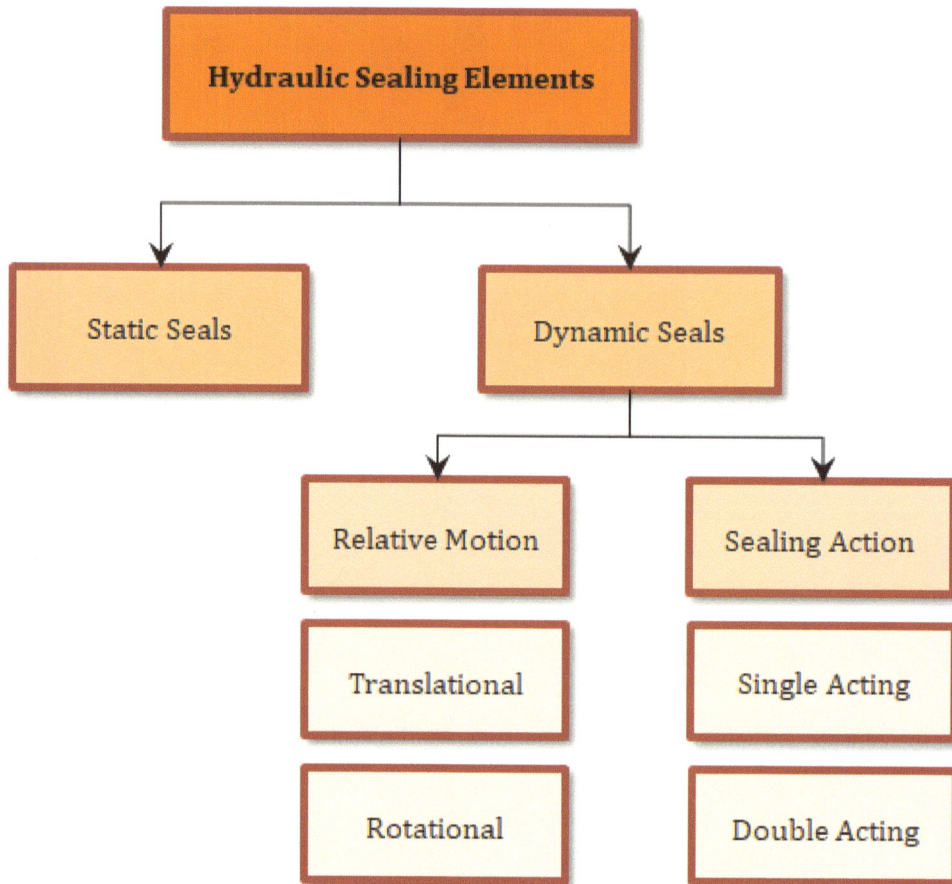

Fig. A.3 - Classifications of Hydraulic Sealing Elements

Static Seals:
- Used to seal between mating surfaces that have no relative motion between them.
- Used to fill confined or none-confined spaces.
- Basic static seals include Sealing Rings and Gaskets.

Dynamic Seals:
- Used to seal and lubricate between surfaces where at least one of them is moving.
- A dynamic seal could be installed on a stationary surface, e.g. Cylinder rod seal.
- A dynamic seal could be installed on a moving surface, e.g. Cylinder piston seal.
- Relative motion between sealed surfaces could be translational as in cylinders and spool valves or could be rotational as in pumps and motors.
- A dynamic seal could seal in one direction or in two directions

Based on specific construction, the following are common sealing configurations:

- Sealing Rings
- Cup Seals
- U-Cup Seals
- T-Shaped Seals
- V-Packings
- Spring-Energized Seals
- Glands
- Wear-Rings
- Back-up Rings
- Rod Wipers (Scrapers)

Chapter 2 – Sealing Rings

2.1- Features and Basic Use of Sealing Rings

Sealing Rings are common sealing elements because they have several advantages:
- They are available in all sizes and all type of elastomers.
- They work over a wide range of operating pressure.
- They are easy to assemble and to replace.
- They require small room to fit inside the components.
- Their failure analysis isn't difficult.
- They are inexpensive and readily available.
- Used for both static and dynamic sealing and for translational and rotational shafts.
- As shown in Fig. A.4, they can be sold as separate pieces with specific dimensions or as *Cord Stock*.

Fig. A.4 - Example of O-Rings Cord Stock (Courtesy of MFP Seals)

As shown in Fig. A.5, *Sealing Rings* are used as static seals against external leakage from between parts in a hydraulic component.

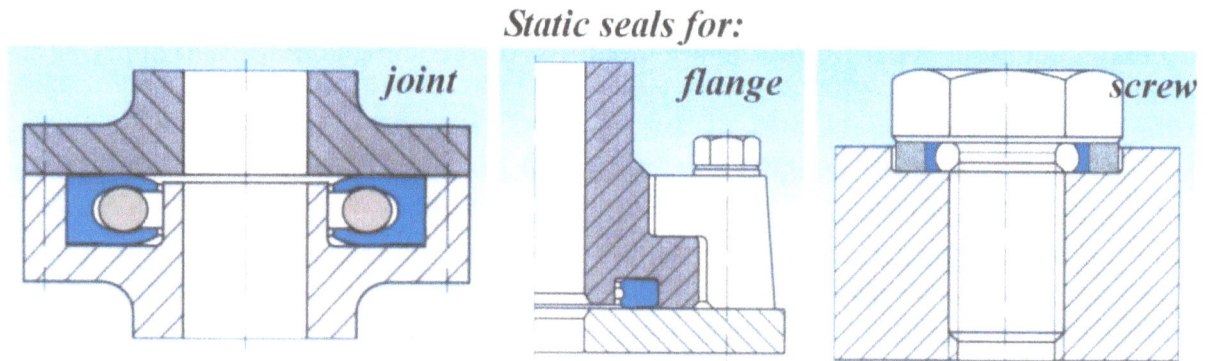

Fig. A.5 - Use of Sealing Rings as Static Seals (Courtesy of ASSOFLUID)

As shown in Fig. A.6, Sealing Rings have limited use as dynamic seals under the following conditions:
- For short-strokes.
- For relatively small diameter applications.
- For relatively low-pressure applications.

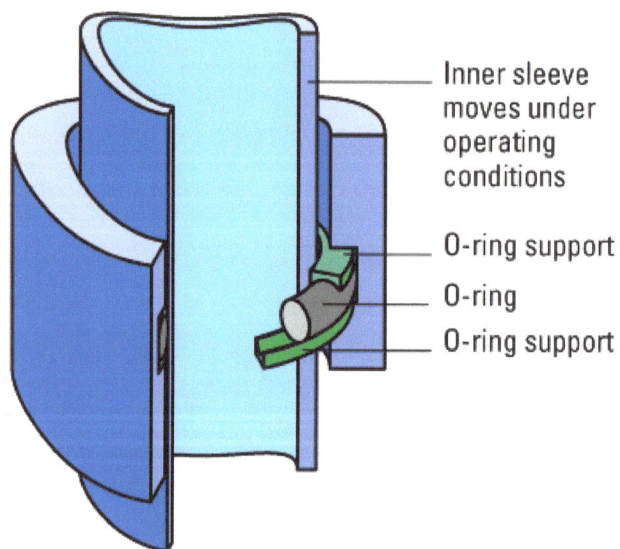

Fig. A.6 - Use of Sealing Rings as Dynamic Seals

2.2- Configurations of Sealing Rings

Sealing Rings have various cross sections as shown in Fig. A.7. Despite the various cross sections, they are typically called O-Rings. In this chapter, O-Ring refers to the ones with circular cross section.

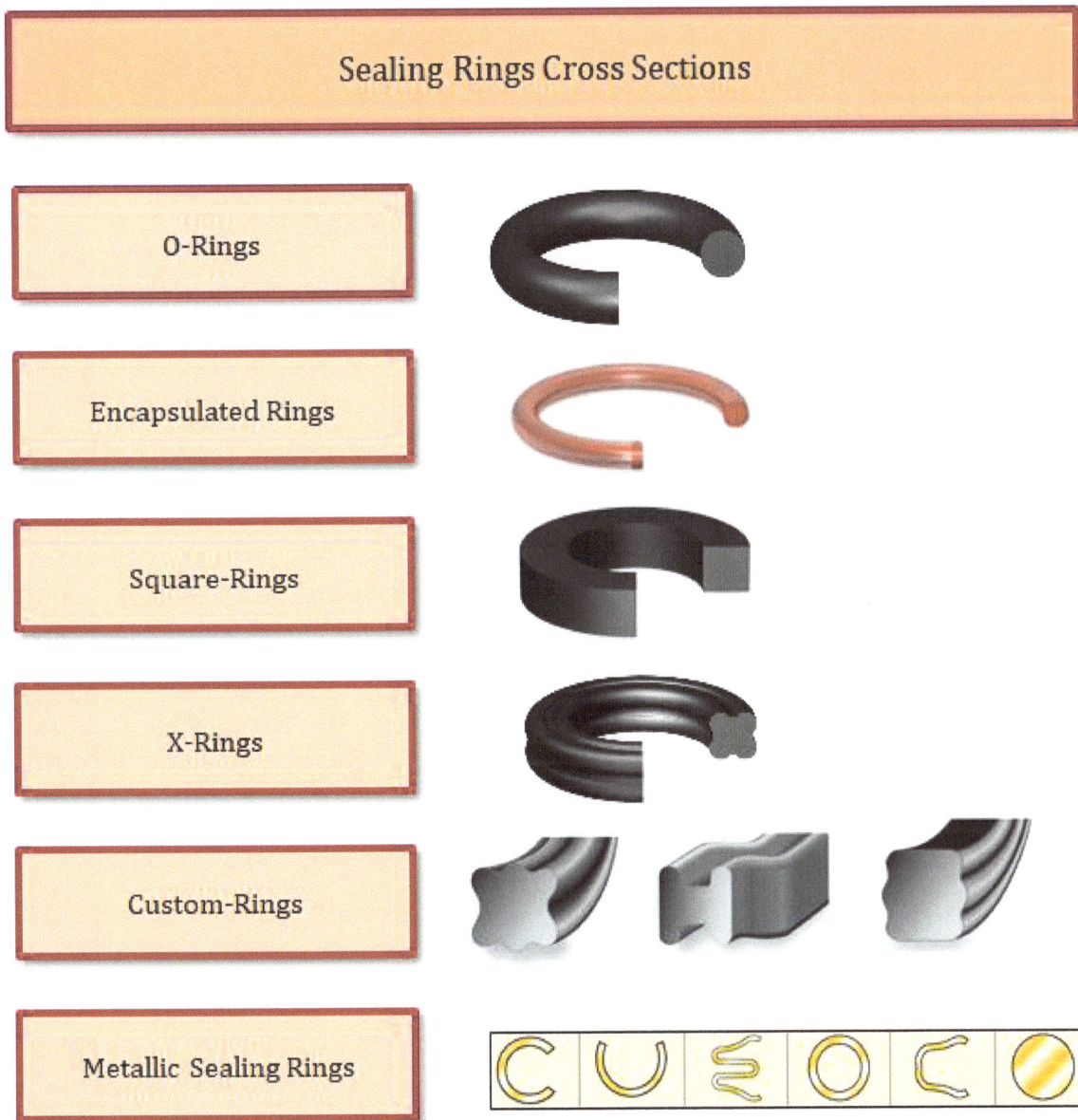

Sealing Rings Cross Sections

- O-Rings
- Encapsulated Rings
- Square-Rings
- X-Rings
- Custom-Rings
- Metallic Sealing Rings

Fig. A.7 - Use of Sealing Rings as Dynamic Seals

2.3- O- Rings

O-Rings were one of the first developed and are the most common type of sealing rings.

2.3.1- O-Rings Construction

As shown in Fig. A.8, an *O-Ring* is a sealing element with solid circular cross section. They are produced in a wide range of materials and dimensions.

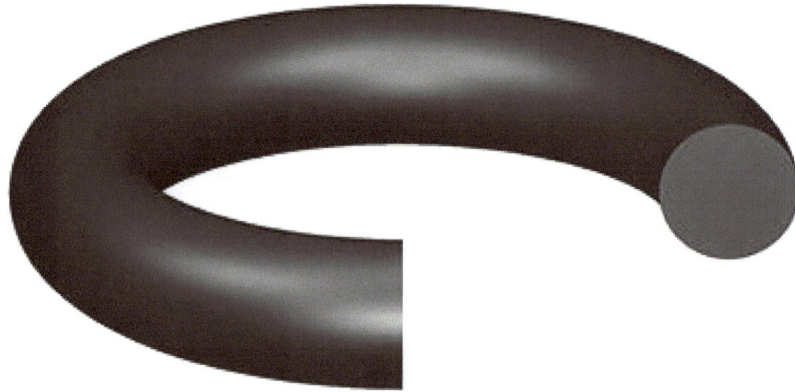

Fig. A.8 - Basic Construction of Static O-Rings (Courtesy of MFP Seals)

2.3.2- O-Rings Sealing Mechanism

As shown in Fig. A.9, an O-Ring is compressed (15-30) % of its initial volume after assembly. The initial squeeze, which acts in a radial or axial direction depending on the installation, gives the O-Ring its initial sealing capability. These forces are superimposed by the system pressure to create the total sealing force which increases as the system pressure increases

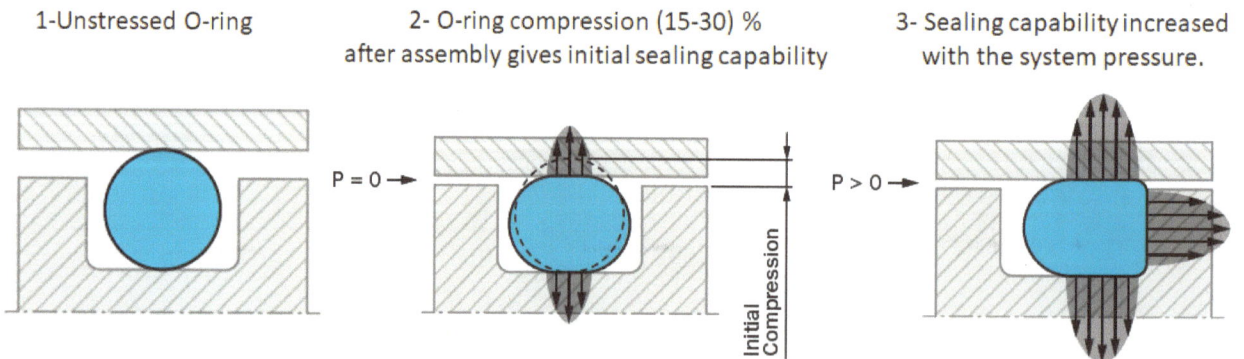

1-Unstressed O-ring

2- O-ring compression (15-30) % after assembly gives initial sealing capability

3- Sealing capability increased with the system pressure.

$P = 0$

$P > 0$

Initial Compression

Fig. A.9 - O-Ring Sealing Mechanism (Courtesy of Trelleborg)

2.3.3- O-Rings Main Dimensions

O-Rings are produced in inch and metric standard sizes and are available from numerous manufacturers. Standard sizes are based on **ISO 3601** and **SAE AS568** standards

Figure A.10 shows the main dimension of O-Rings as follows:
- Inside Diameter (ID).
- Radial and Axial Crosssection (W).
- Radial and axial Flashes due to rubber injection.

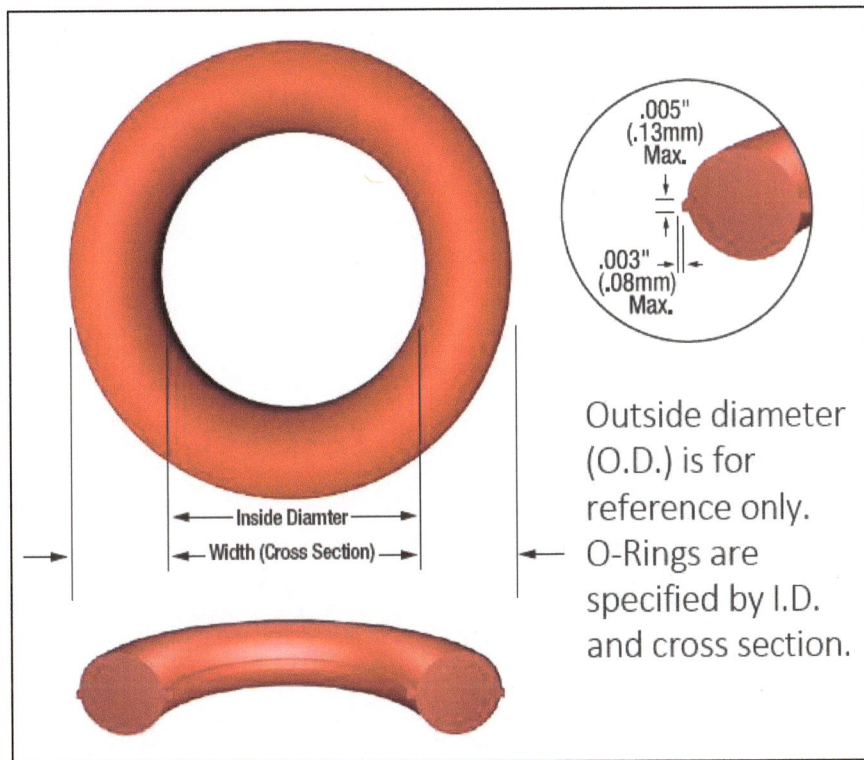

.005"
(.13mm)
Max.

.003"
(.08mm)
Max.

Inside Diamter

Width (Cross Section)

Outside diameter (O.D.) is for reference only. O-Rings are specified by I.D. and cross section.

Fig. A.10 - Main Dimensions of Basic O-Rings (www.applerubber.com)

2.4- Encapsulated O-Rings

Figure A.11 shows the construction of an *Encapsulated O-Ring.* It consists of an elastomeric base material that is encapsulated with *Teflon* industrial coating (FEP). The FEP coating provides increased chemical, temperature and wear resistance that would otherwise not be achievable with the base compound alone. Figure A.12 shows typical examples of construction of encapsulated O-Rings.

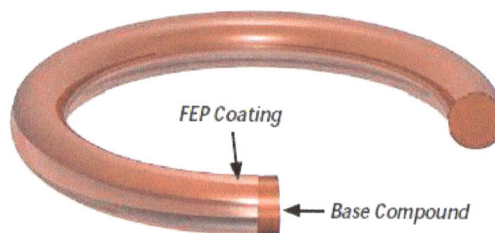

FEP Coating

Base Compound

Fig. A.11 - Construction of Encapsulated O-Rings (Courtesy of MFP Seals)

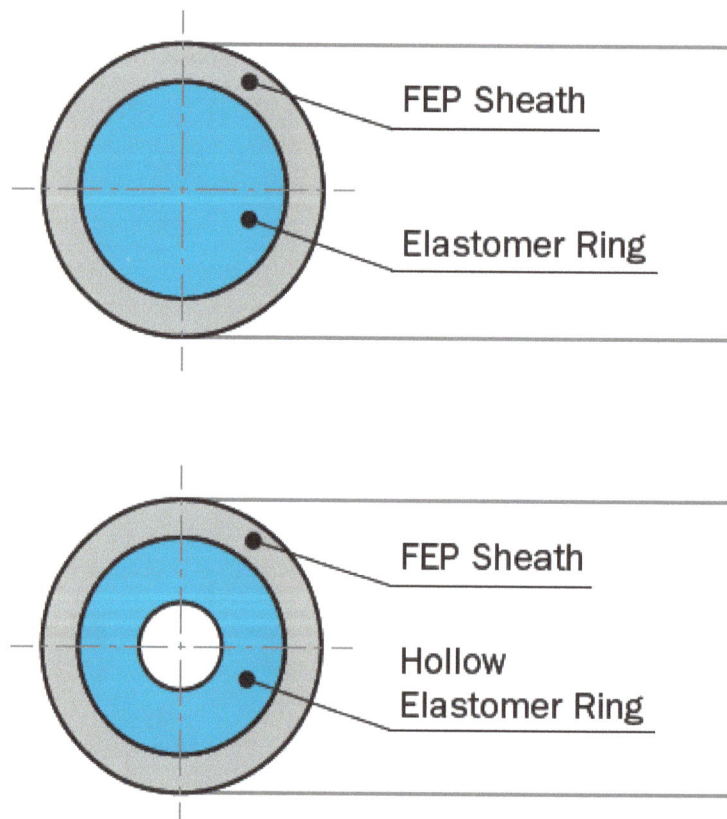

FEP Sheath

Elastomer Ring

FEP Sheath

Hollow
Elastomer Ring

Fig. A.12 - Typical Construction of Encapsulated O-Rings (Courtesy of Trelleborg)

2.5- Square-Rings

Figure A.13 shows the construction of *Square-Rings*. They are used for the same function as O-Rings. They are directly interchangeable in certain sizes with the O-Rings, using the same groove. Due to their larger sealing surface, Square-Rings can normally handle higher pressures than O-Rings. Square-Rings are commonly used for static applications. Figure A.14 shows the main dimensions of Square-Rings.

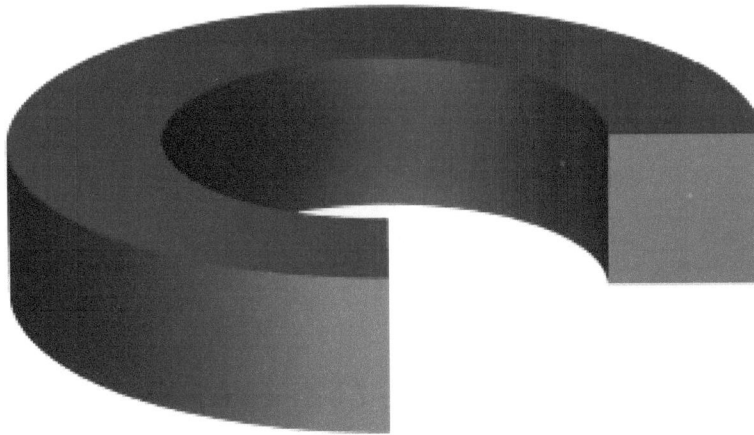

Fig. A.13 - Construction of Square-Rings (Courtesy of MFP Seals)

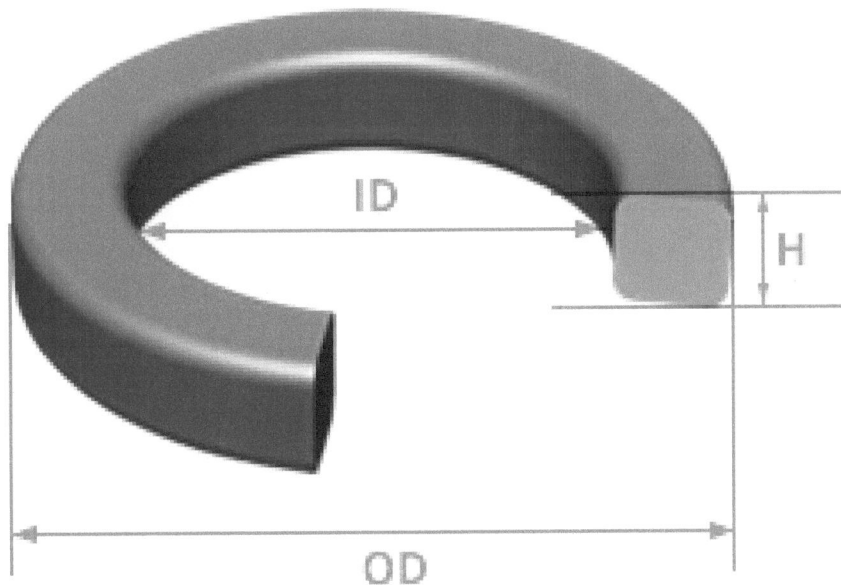

Fig. A.14 - Main Dimensions of Square-Rings (www.marcorubber.com)

2.6- X- Rings

Figure A.15 shows the construction of *X-Rings* that are also referred to as *Quad-Rings*. They are designed to offer two contact points on each sealing surface versus one contact point in O-Rings. This dual sealing lip feature gives positive sealing with reduced radial squeeze, and thus reduced friction. By reducing the friction, the wear of ring is reduced giving the seal a longer service life over the standard O-Ring design. X-Rings are excellent for use in rotary applications and for providing seal stability due to a design that resists twisting in the groove. Figure A.16 shows the main dimension of X-Rings.

Fig. A.15 - Construction of X-Rings (Courtesy of MFP Seals)

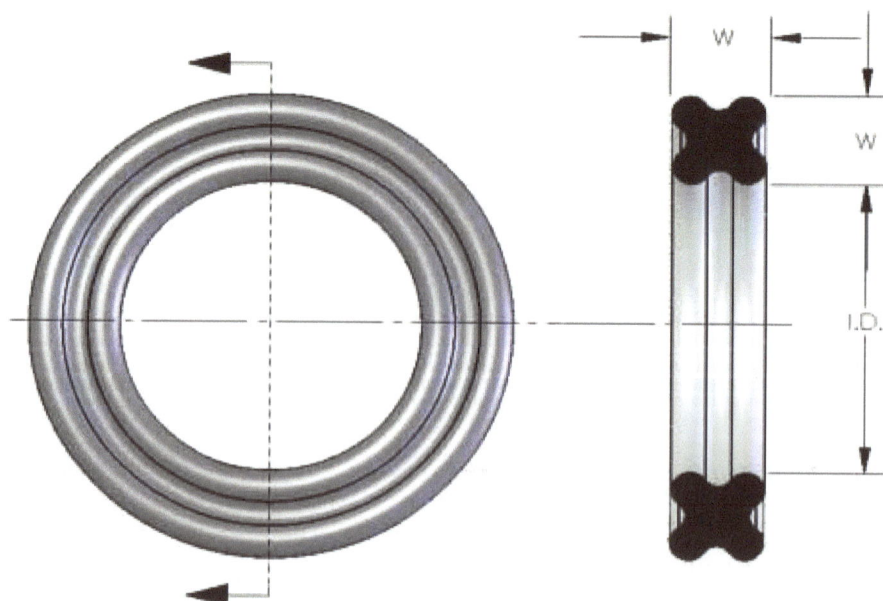

Fig. A.16 - Main Dimensions of X-Rings (www.marcorubber.com)

2.7- Custom-Rings

Figure A.17 shows the construction of a *Custom-Ring,* example 1. Such a ring is designed for dynamic sealing applications providing near zero leakage at pressures up to 138 bar (2000 psi). This six-lobed configuration, designed with two primary and four backup sealing surfaces, has excellent sealing features in very difficult applications. It can be used with standard O-Ring grooves.

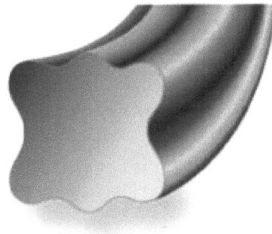

Fig. A.17 - Quad-O-Dyn® Brand Seals (www.mnrubber.com)

Figure A.18 shows the construction of a *Custom-Ring,* example 2. Such a ring is ideal to fill single or multiple groove configurations in static face seal applications.

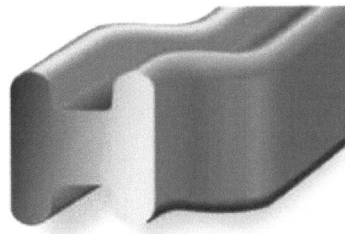

Fig. A.18 - Quad®-O-Stat Brand Seals (www.mnrubber.com)

Figure A.19 shows the construction of another *Custom-Ring,* example 3. Such a ring is designed specifically for static face sealing applications. Each of the six contact points serves as an individual seal with the corner lobes functioning as seal backups to the central lobes. If one lobe fails, the remaining lobes provide zero leakage sealing. They can be installed in standard O-Ring grooves.

Fig. A.19 - Quad® P.E. Plus Brand Seals (www.mnrubber.com)

2.8- Metallic Sealing Rings

Cross Sections: As shown in Fig. A.20, *Metallic Sealing Rings* are available from various manufacturers in a variety of cross sections. Like the rubber-based seals, they could be non-spring energized, or spring energized.

Material: Materials include Steel, Stainless Steel, Cast Iron, Ductile Iron and Bronze. Piston ring coatings of Manganese Phosphate, Tin Nickel, PTFE/Nickel, Chrome and Silver Lead Indium are available to maximize performance for lubricating requirements, corrosion resistance, wear, and friction reduction.

Fluid Compatibility: They are compatible with petroleum base and synthetic fluids and phosphate esters among others.

Major Advantages:
- Reliable under severe operating conditions for extended periods of time.
- High pressure sealing up-to 25,000 psi (1700 bar) without the risk of blow-by.
- Service temperature up to 1800 °F (982 °C).

Selection: The figure shows the selection criteria based on the operating conditions.

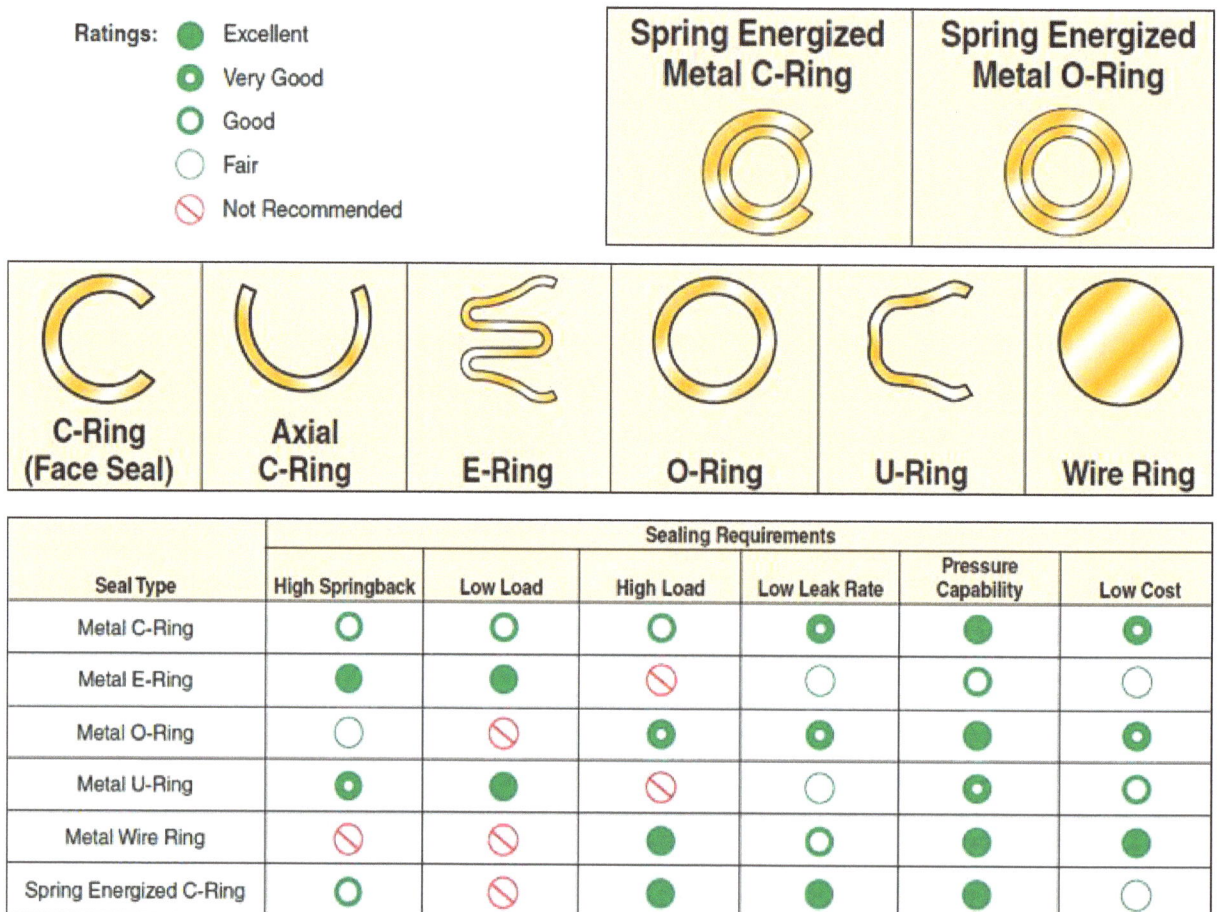

Ratings:
- ● Excellent
- ◉ Very Good
- ○ Good
- ○ Fair
- ⊘ Not Recommended

Seal Type	Sealing Requirements					
	High Springback	Low Load	High Load	Low Leak Rate	Pressure Capability	Low Cost
Metal C-Ring	Good	Good	Good	Very Good	Excellent	Very Good
Metal E-Ring	Excellent	Excellent	Not Recommended	Fair	Good	Fair
Metal O-Ring	Good	Not Recommended	Very Good	Very Good	Excellent	Very Good
Metal U-Ring	Very Good	Excellent	Not Recommended	Fair	Very Good	Good
Metal Wire Ring	Not Recommended	Not Recommended	Excellent	Good	Excellent	Excellent
Spring Energized C-Ring	Good	Not Recommended	Excellent	Excellent	Excellent	Good

Fig. A.20 - Metallic Sealing Rings (Courtesy of Parker)

Chapter 3- Hydraulic Seals

3.1- Cup Seals

Figure A.21 shows the basic *Cup Seal* shape. Such a seal performs the sealing function in one direction where the pressure increases the sealing force. This cup seal requires a backing plate to retain it in position and is used for pressure up to 70 bar (1,000 psi).

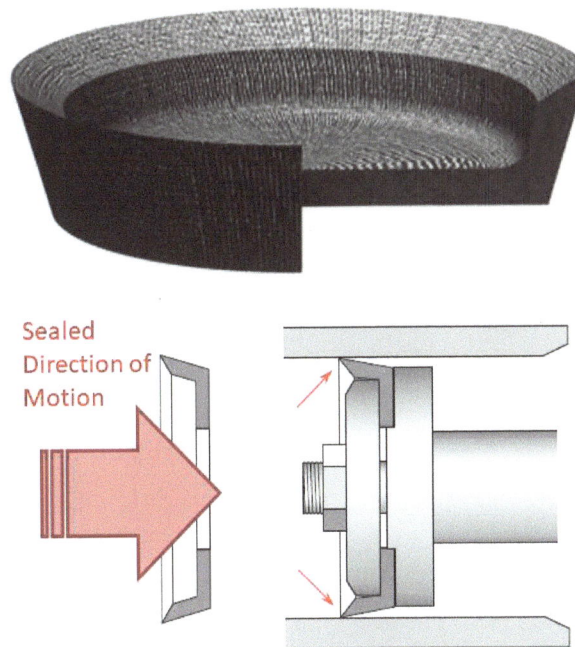

Sealed
Direction of
Motion

Fig. A.21 - Cup Seals (Curtesy of MFP Seals)

3.2- U- Cup Seals

Figure A.22 shows an asymmetric *U-Cup Seal*. U-Cups are very popular in industrial cylinders and can be used as piston or rod seals. When used with back-up rings, system pressure should not exceed 173 bar (2,500 psi).

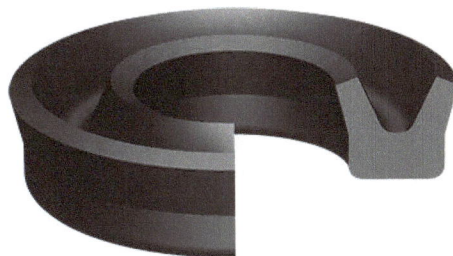

Fig. A.22 - U-Cup Seal (Courtesy of MFP Seals)

Figure A.23 shows the sealing mechanism of a U-Cup seal. As shown in the figure, the U-Cup Seal performs both static and dynamic sealing. It consists of a static sealing lip, and a dynamic sealing lip. Like Cup seals, it seals in one direction and relaxes in the other.

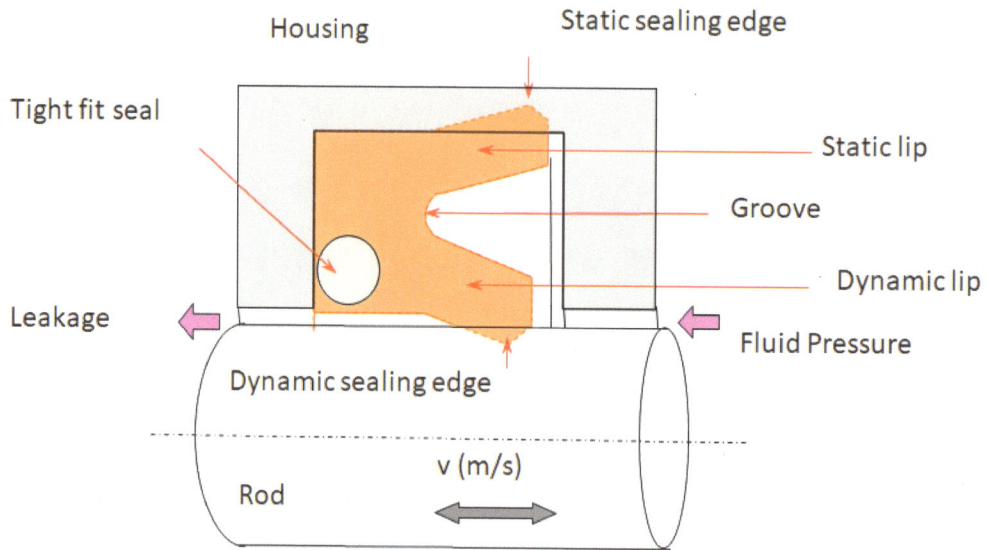

Fig. A.23 - U-Cup Sealing Mechanism

Figure A.24 shows a symmetric O-Ring loaded squeeze U-Cup seal with and without anti-extrusion Back-up Ring. Back-up rings are available and required for use with pressure up to 400 bar (5,800 psi) pressure.

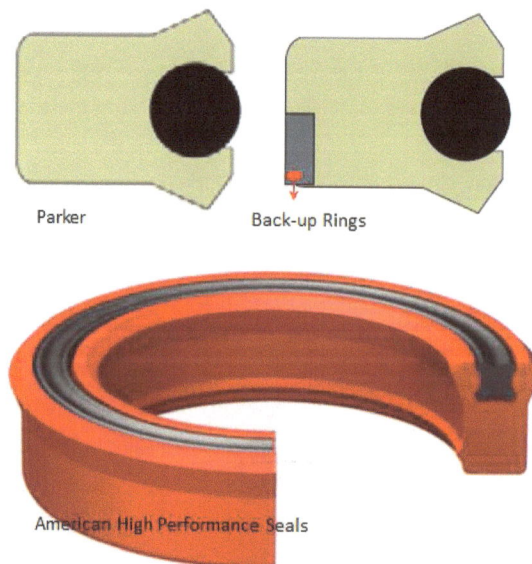

Fig. A.24 - Symmetric U-Cup Seals (Courtesy of Parker)

3.3- T- Shaped Seals

Figure A.25 shows various types of *T-Shaped* seals. They are used for sealing both cylinder rods and pistons for pressure up to 700 bar (1,0152 psi). Each of the shown below sealing packages consists of a T-Shaped rubber sealing ring and the two anti-extrusion Back-up Rings.

Cylinder Rod T-Shaped Seals   Cylinder Piston T-Shaped Seals

Fig. A.25 -T-Shaped Seals (Courtesy of American High-Performance Seals)

Figure A.26 shows a T-Shaped seal produced by different manufacturer. The use of the supporting Back-up Rings on both sides of the seal eliminates rolling or spiraling in long stroke cylinders or dry rod conditions.

Fig. A.26 - T-Shaped Sealing Package (Courtesy of MFP Seals)

3.4- V-Packings

The *V-Packing* is a multi-part sealing set. They are used for sealing both cylinder rods and pistons. The V-Packing, as shown in Fig. A.27, consists of a one (top) female adaptor and a one(bottom) male adaptor, and V-Rings between them. The female adaptor is also referred to as *Backup Ring* or *Base Ring*. It is manufactured from an elastomer with good extrusion resistance. The male adaptor is also referred to as *Compression Ring* or *Energizing Ring*. It ensures the uniform loading pressure distribution on the other rings. The number of the V-Rings depends on the operating pressure. The material of the V-Rings depends on the application. As shown in Fig. A.28, V-Rings of different material can be used in the same pack to get the best performance. This type of seal is usually used for high pressure applications.

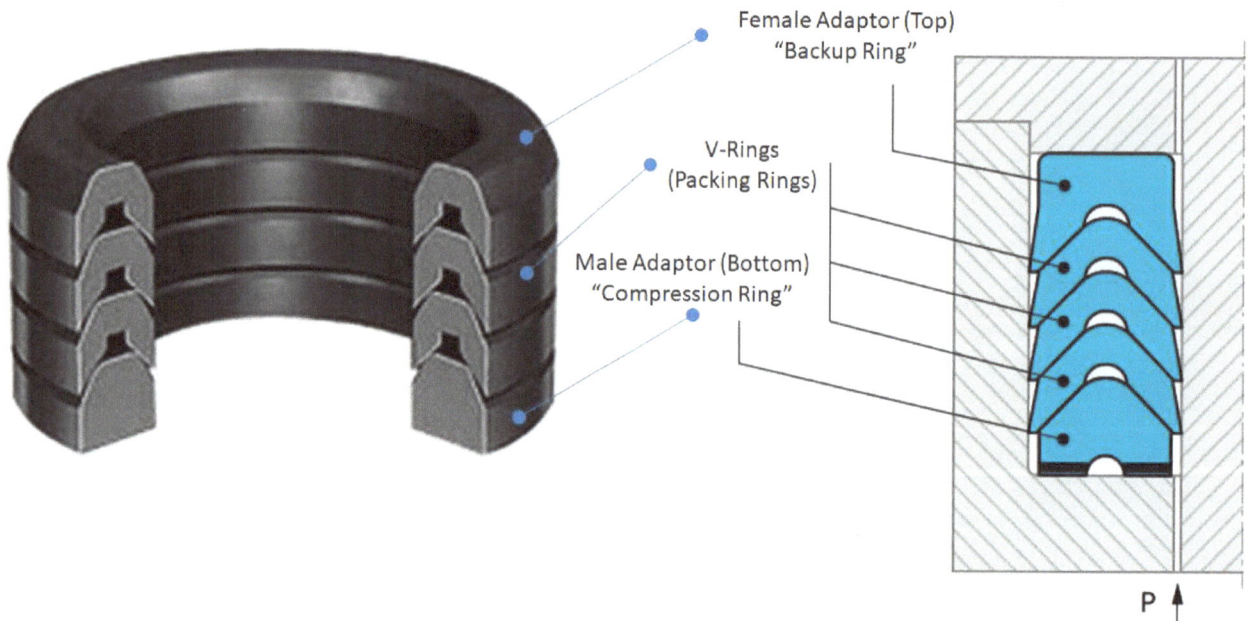

Female Adaptor (Top)
"Backup Ring"

V-Rings
(Packing Rings)

Male Adaptor (Bottom)
"Compression Ring"

P

Fig. A.27 - V-Packings (Courtesy of Trelleborg)

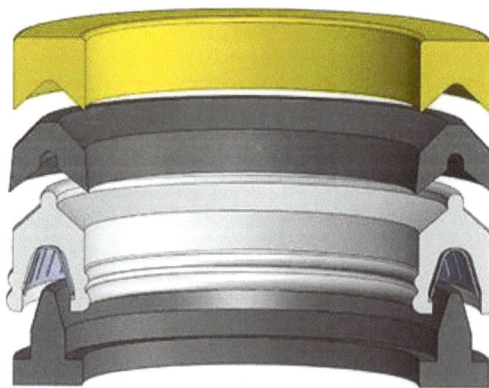

Fig. A.28 - V-Packings of Different Materials (www.hydrapakseals.com)

3.5- Spring-Energized Seals

Figure A.29 shows various configurations of *Spring-Energized* seals. They are used for static and dynamic sealing. Spring-Energized seals are more dynamically stable and typically used in areas where elastomeric seals cannot meet the frictional, temperature, pressure, or chemical-resistance requirements of the application.

Common uses for Spring-Energized Seals are:
- Explosive decompression resistant applications.
- Applications with extreme operating temperature and/or pressure.
- Application with high surface speeds.
- Non-lubricated applications.

Fig. A.29 - Spring-Energized Seals (Courtesy of MFP Seals)

Figure A.30 shows a typical example of a Spring-Energized Plastic U-Cup seal with the following advantages:

- Suitable for reciprocating and rotary applications.
- Low coefficient of friction.
- Stick-slip free operation.
- High abrasion resistance.
- Dimensionally stable.
- Resistant to most fluids, chemicals and gases.
- Withstands rapid changes in temperature.
- Excellent resistance to aging.
- Interchangeable with O-Ring and Back-up Ring combinations.

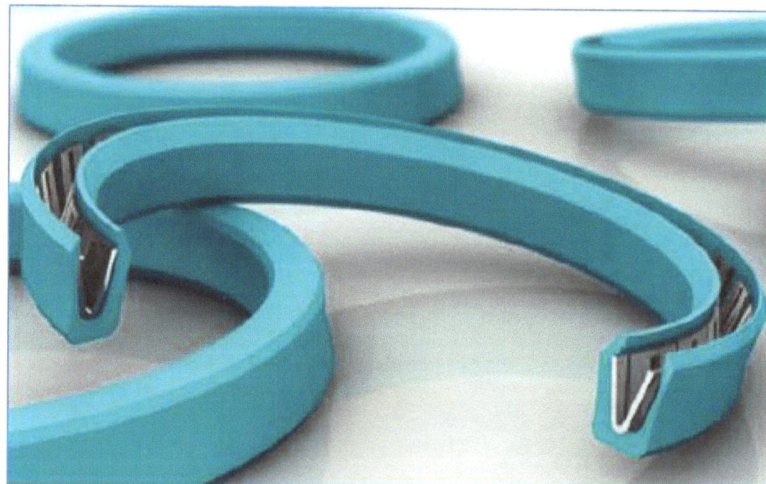

OPERATING CONDITIONS

Pressure:	Maximum dynamic load: 20 MPa Maximum static load: 40 MPa (207 MPa with back-up ring)
Speed:	Reciprocating up to 15 m/s Rotating up to 1.27 m/s
Operating temperature:	-70 °C to +300 °C Special Turcon and Zurcon® materials as well as alternative spring materials are available for applications outside this temperature range.
Media compatibility:	Virtually all fluids, chemicals and gases

Fig. A.30 - Spring-Energized U-Cup Seals (Courtesy of Trelleborg)

Figure A.31 shows the sealing mechanism for the seals shown in the previous figure. The spring supplies the load required for sealing at low pressures. Fluid pressure energizes the sealing lips, so total sealing force rises with increasing operating pressure.

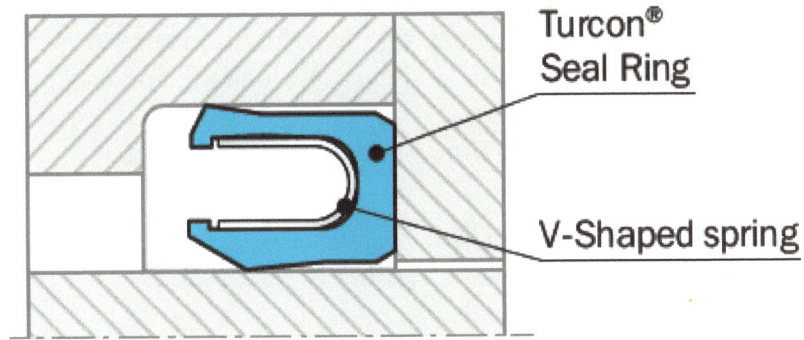

Turcon® Seal Ring

V-Shaped spring

Spring Force without system pressure

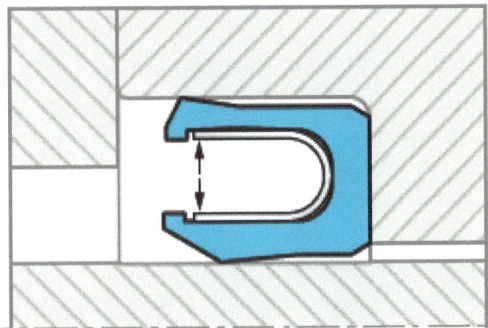

Sealing Force after applying system pressure

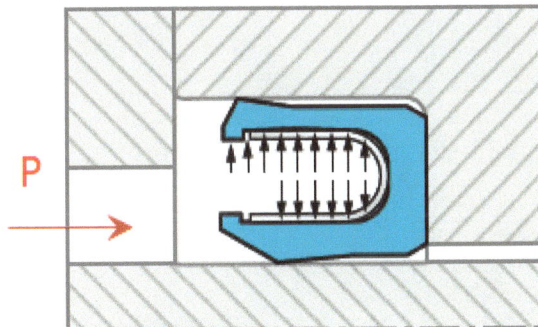

P

Fig. A.31 - Sealing Mechanism of a Typical Spring-Energized Seal (Courtesy of Trelleborg)

3.6- Wear-Rings

Hydraulic seals are not designed to provide bearing surfaces or to carry lateral loads. As shown in Fig. A.32, the function of *Wear-Rings* is to:
- Provide bearing surfaces to absorb side loads of cylinder rods and/or pistons.
- Prevent metal-to-metal contact that would otherwise damage and score the sliding surfaces and eventually cause seal damage, leakage and component failure.

Wear-Rings are also referred to as *Guide-Rings* or *Wear-Bands*.

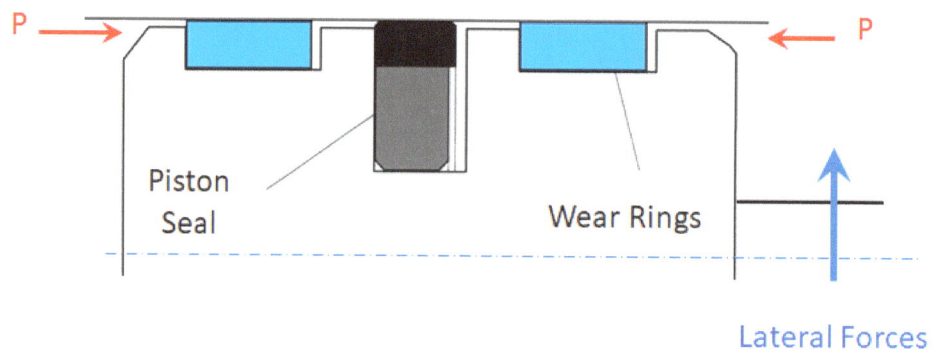

Fig. A.32 - Wear-Rings

Figure A.33 shows some typical Wear-Rings that are made of different material such as Nylon, Teflon and other plastics. Like O-Rings, as shown in Fig. A.34, Wear Rings can be delivered as pieces or strips.

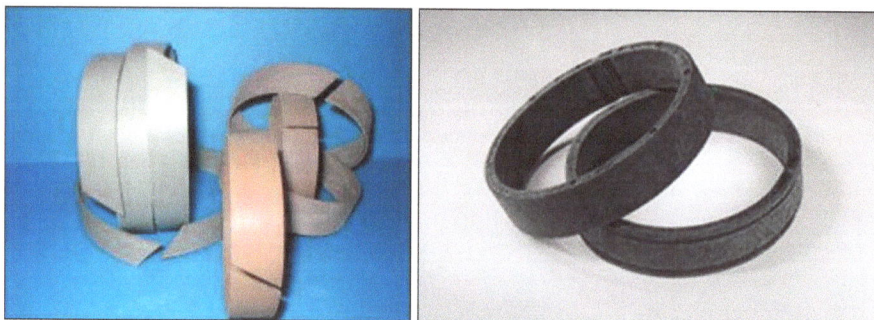

Fig. A.33 - Examples of Wear Rings

Fig. A.34 - Wear-Strip (Courtesy of MFP Seals)

Wear Rings are available, as shown in Fig. A.35, with various cut designs. Figure A.36 shows a Wear-Ring design with a special texture on the sliding surface. It is named *TEARDROP* profile. Such a profile has small lubricant pockets on the surface which improve the initial lubrication and promotes the formation of a lubricant film. They also help to protect the seal system through their ability to embed any foreign particles.

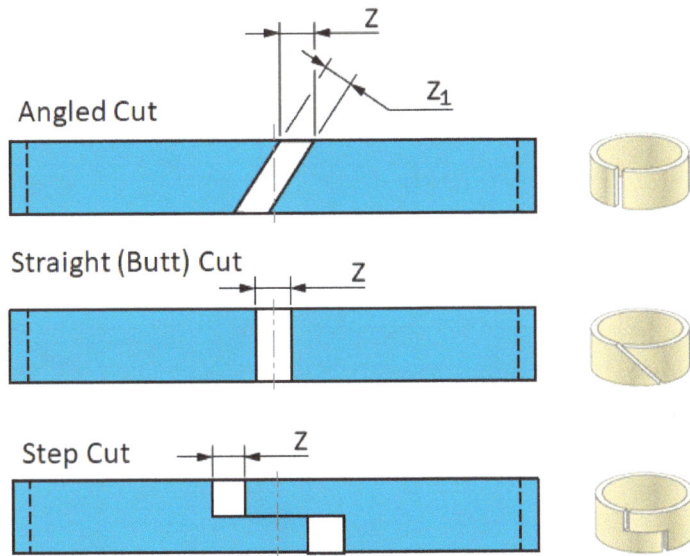

Fig. A.35 - Wear-Rings Cut Design (Courtesy of Trelleborg)

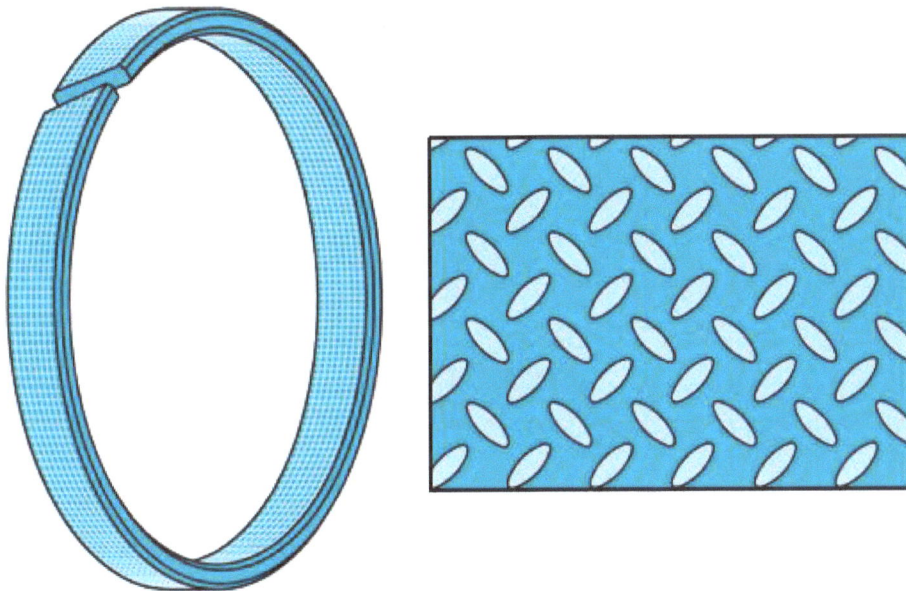

Fig. A.36 - Wear-Rings with TEARDRP Profile on the Sliding Surface (Courtesy of Trelleborg)

Equation A.1 shows how to calculate the minimum bearing length.

$$P_R = \frac{F \times f}{\textbf{Projected Area}} = \frac{F \times f}{D \times T} \rightarrow T = \frac{F \times f}{D \times P_R}$$

A. 1

Where (as shown in Fig. A.37):

- P_R = Design Pressure for a specific Wear-Ring.
- F = maximum estimated lateral force based on cylinder design and application.
- f = Safety factor (default = 2).
- D = Diameter of cylinder rod or piston.
- T = minimum bearing length.

Example: A wear ring is required for a cylinder rod of 60 mm diameter and 40,000 N maximum estimated radial force. A specific Wear-Ring material is selected that has 100 N/mm² design pressure. Then the minimum width is:

$$T \text{ (mm)} = \frac{40.000 \times 2}{60 \times 100} = 13.3$$

Assuming the manufacturer design tables shows Wear-Rings with a maximum length of 10 mm, installation of two rings is recommended to provide a longer guide length.

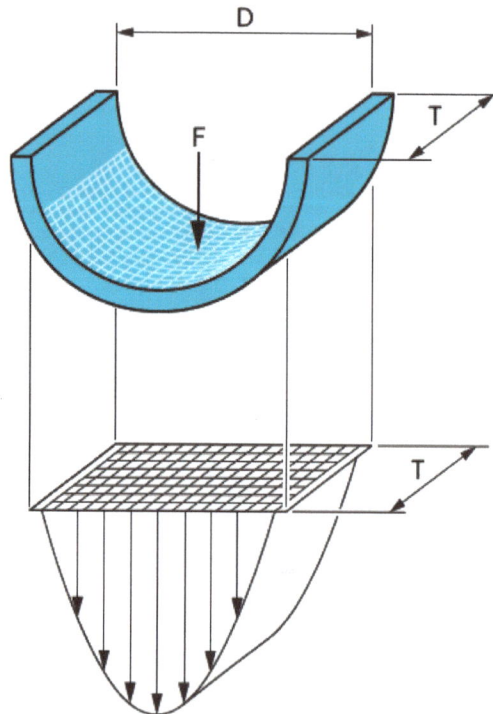

Fig. A.37 - Calculation of the Bearing Length (Courtesy of Trelleborg)

Figure A.38 shows typical examples of Wear-Rings with various cross sections that are characterized by:

- High load bearing capabilities.
- High operating temperatures and pressures.
- Cost effective and long service life.
- Easy installation and replacement.
- Wear-resistant and long service life.
- Low friction and self-lubrication.
- Wiping/cleaning effect.
- Ability to embed foreign particles possible.
- Damping of mechanical vibrations.

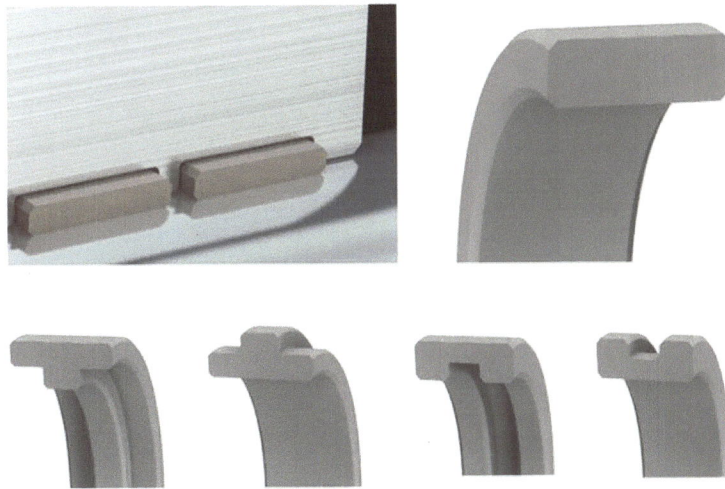

Fig. A.38 - Wear-Rings of Various Cross Sections (Courtesy of American High-Performance Seals)

Figure A.39 shows a typical example of Wear-Rings assembled on a cylinder piston (left) and in a cylinder rod glands (right).

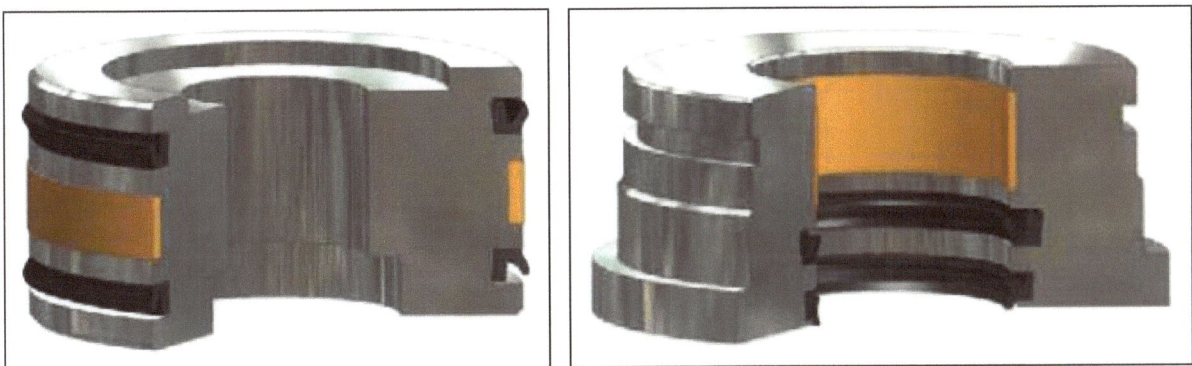

Fig. A.39 - Typical Wear-Rings for a Cylinder Piston and Rod Gland (Courtesy of Parker)

3.7- Backup Rings

The main purpose of standard *Backup Rings* is to prevent seal extrusion Backup rings are produced in a variety of materials such as leather, rubber, elastomers and Teflon. Figure A.40 shows various types of the standard Backup Rings, each of which is produced with a wide range of dimensions. Figure A.41 shows examples of sealing rings supported by Backup Rings. Figure A.42 shows a typical example of Back-up Rings assembled in a cylinder piston (left) and in a cylinder rod gland (right).

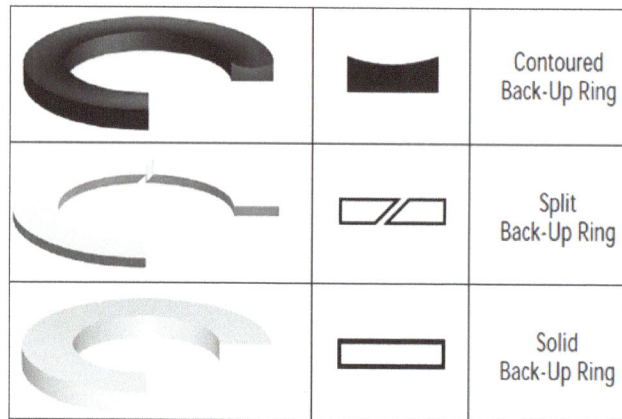

Fig. A.40 – Cross Sections Standard Backup Rings (Courtesy of MFP Seals)

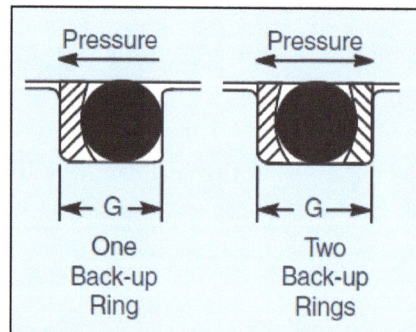

Fig. A.41 - Examples of Sealing Rings Supported by Backup Rings (Courtesy of Parker)

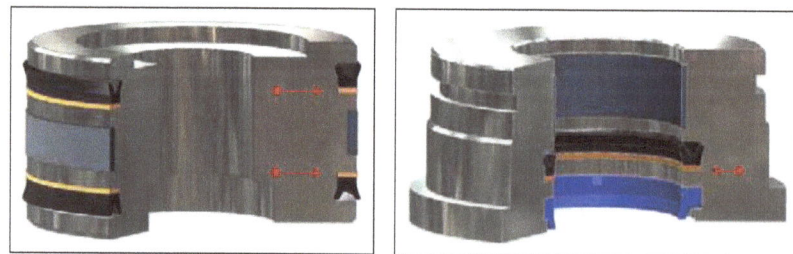

Fig. A.42 - Typical Back-up Rings for a Cylinder Piston and Rod Gland (Courtesy of Parker)

3.8- Rod Wipers

Contaminated hydraulic fluid is a frequent cause of total system failure. As a result of increasingly sensitive elements within the hydraulic system, the Rod-Wipers function becomes ever more important.

Rod *Wipers* (also known as *Scrapers*, *Excluders* or *Dust Seals*) are mainly used as:
- The first line of defense against contamination that has settled on the cylinder rod.
- The last line of defense against external leaking.
- Host lubrication film enough for continuous self-lubrication of the cylinder rod.

As shown in Fig. A.43, a traditional Rod Wiper is referred to as a single-acting Wiper because it prevents contamination from getting into the cylinder during rod retraction. It performs both static and dynamic sealing. They are usually made from a mix of brass and some other material to compromise the required hardness and the flexibility. Figure A.44 shows a typical example of a basic wiper seal.

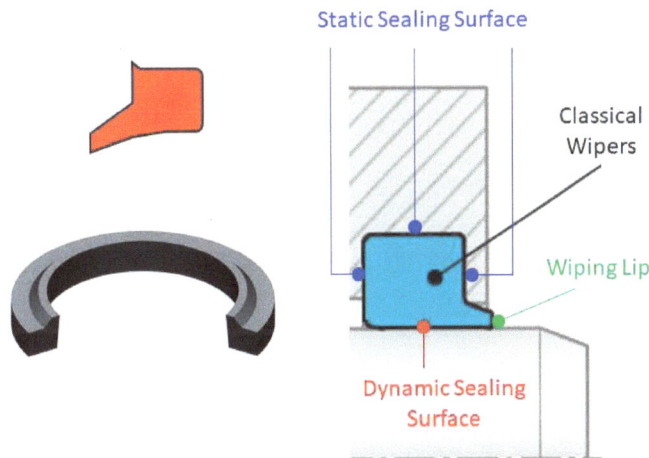

Fig. A.43 - Traditional Single-Acting Rod Wiper Seal

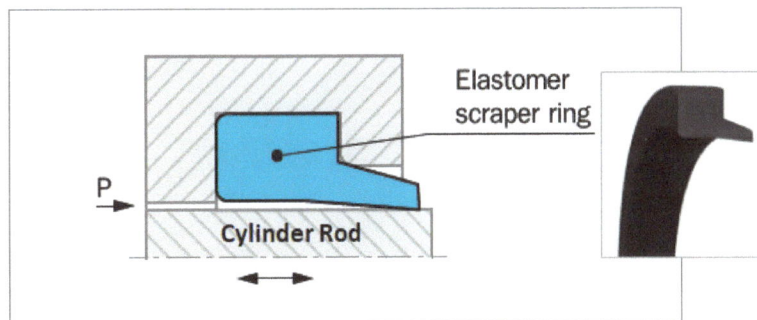

Fig. A.44 - Typical Basic Single-Acting Rod Wiper Seal (Courtesy of Trelleborg)

They are usually made from a mix of brass and some other material to achieve the required hardness and the flexibility. Modern Rod Wiper Seals may contain more than one scraper lip.

Example 1: Figure A.45 shows a typical example of a Single-Acting Rod Wiper Seal that contains *Redundant* sealing lips, a thin metallic lip and an elastomeric lip. The two scraper lips are engaged in tandem behind each other in a compact metal housing. The metallic lip is self-adjusting, lasts longer, and effectively removes contaminants from the rod surface. Rubber lips are the last line of defense against contaminants and provide better sealing.

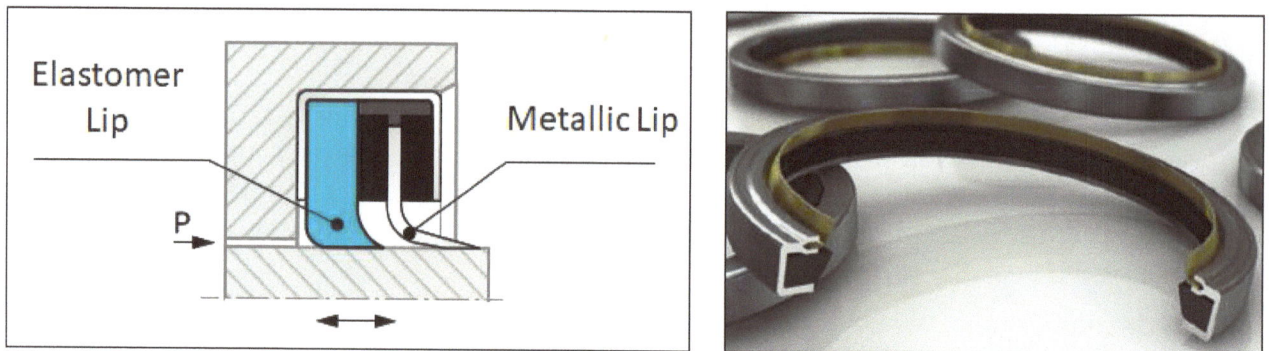

Fig. A.45 - Single-Acting Wipers with Redundant Sealing Lips (Courtesy of Trelleborg)

Example 2: Figure A.46 shows a typical example of a *Double-Acting Wiper Seal*. The scraper lip is designed in a particular way that it reliably scrapes off the dirt but leaves a residual oil film on the rod, which is required for correct operation. The radial squeeze is sufficient to remove particles, dust and water. The scraping lip facing inwards is designed in a way that it assumes a sealing function even under low pressure. The static seal is achieved by a tight radial fit between the scraper body and the groove.

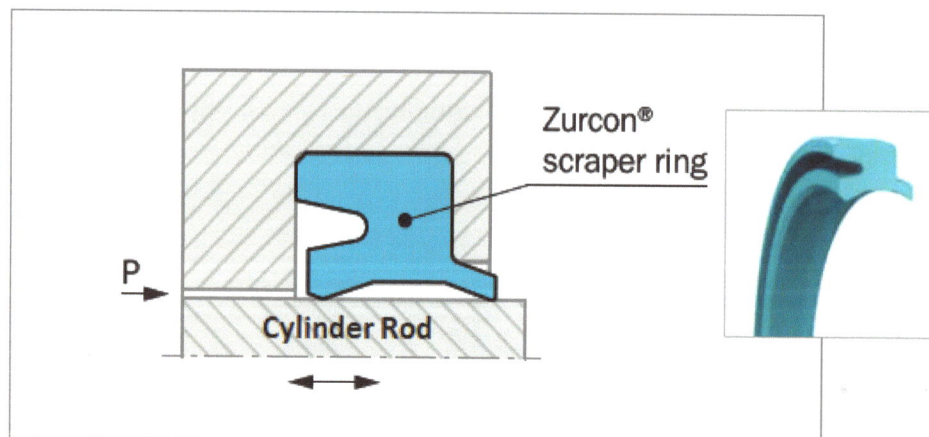

Fig. A.46 - Double-Acting with Redundant Sealing Lips (Courtesy of Trelleborg)

Example 3: Figure A.47 shows a high-performance rod wiper referred to as *"Umbrella Wiper Technology"*. Such wipers are designed for harsh environment where the wiper is subjected to numerous contaminants. What makes this design unique is the protective guard that entirely covers the retaining groove. Typical applications are agriculture, off-highway, and forestry equipment.

Advantages of this type of wiper are as follows:

- Directs water and debris away from rod and housing.
- Prevents traditional ingression of moisture and mud.
- Eliminate corrosion in wiper and gland housing
- Unlike rod wipers with metallic lips normally specified for harsh environment, it can be installed without special tools.
- Made from material with good resistance to UV and chemical.
- Offers long life and sealability.

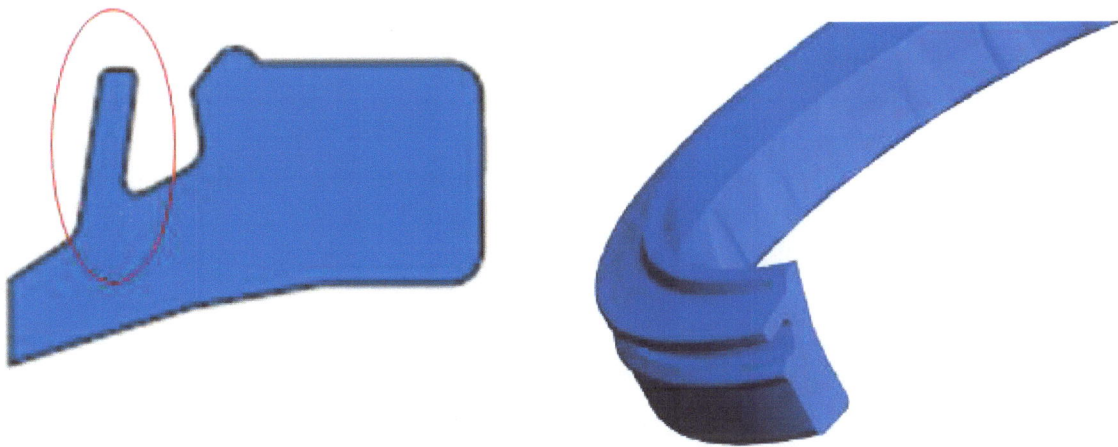

Fig. A.47 - Umbrella Wiper Technology (Courtesy of Hallite Seals)

Chapter 4- Materials for Hydraulic Sealing Elements

Seal materials play a major role in the performance and service life of hydraulic seals, and consequently the reliability of hydraulic components. Generally, hydraulic seals work for a variety of applications and working conditions, such as a wide range of temperature and pressure, contact with various hydraulic fluids, and exposure to the outside environment and side loads. A wide variety of hydraulic seal materials have been developed and used by seal manufacturers.

Like hydraulic components, seal design and performance prediction are based on modeling and simulation. Seal material development is a continuous process that depends on experimentation and field testing. The seal could be produced from one or combinations of the following materials to optimize seal performance:
- Fabric.
- Rubber (natural and synthetic).
- Leather.
- Metal.
- Elastomeric Compounds (a mixture of base polymer and other chemicals that form a finished rubber material).
- Engineered Plastics.

A hydraulic seal can also be composed of a base compound and coated by industrial coating. As shown in Fig. A.48, the most common coating is Polytetrafluoroethylene (PTFE). PTFE coating offer low frictional motion. Consequently, it eliminates sticking, eases seal installation, and minimize power losses.

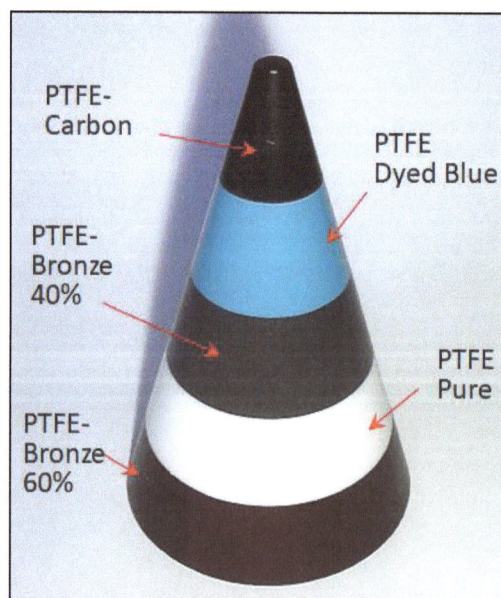

Fig. A.48 - PTFE Seal Material

Every manufacturer has their own codes for seal materials. However, Table A.1, shows the standard abbreviations for synthetic rubbers used in hydraulic seals manufacturing.

ELASTOMER RUBBER COMPOUNDSAND REFERENCES					
General Description	**Chemical Description**	**Abbreviation (ASTM 1418)**	**ISO/DIN 1629**	**Other Trade Names & Abbreviations**	**ASTM D2000 Designation**
Nitrile	Acrylonitrile-butadiene rubber	NBR	NBR	Buna-N	BF, BG, BK, CH
Hydrogenated Nitrile	Hydrogenated Acrylonitrile-butadiene rubber	HNBR	(HNBR)	HNBR, HSN	DH
Ethylene-Propylene	Ethylene propylene diene rubber	EPDM	EPDM	EP, EPT, EPR	BA, CA, DA
Fluorocarbon	Fluorocarbon Rubber	FKM	FPM	Viton ®, Fluorel ®	HK
Chloroprene	Chloroprene rubber	CR	CR	Neoprene	BC, BE
Silicone	Silicone rubber	VMQ	VMQ	PVMQ FC, FE, GE	FC, FE, GE
Fluor-silicone	Fluor-silicone rubber	FVMQ	FVMQ	FVMQ	FK
Polyacrylate	Polyacrylate rubber	ACM	ACM	ACM	EH
Ethylene Acrylic	Ethylene Acrylic rubber	AEM	AEM	Vamac ®	EE, EF, EG, EA
Styrene-butadiene	Styrene-butadiene rubber	SBR	SBR	SBR	AA, BA
Polyurethane	Polyester urethane / Polyether urethane	AU / EU	AU / EU	AU / EU	BG
Natural rubber Natural rubber	Natural rubber Natural rubber	NR	NR	NR	AA

Vamac ® and Viton ® are registered trademarks of E. I. du Pont de Nemours and Company or affiliates.
Fluorel ® is a registered trademark of Dyneon LLC

Table A.1 - Standard Abbreviations for Synthetic Rubber (news.ewmfg.com)

Chapter 5- Properties and Test Methods for Hydraulic Sealing Elements

Nowadays machines are getting faster, and the operating conditions are becoming more severe. This increases the demand for seal material development and testing. The operation of hydraulic seals combines several disciplines such as physics, mechanics, thermodynamics, fluid dynamics, tribology, etc. There are standard test procedures for conducting most of the tests on elastomers. This section overviews the main properties, shown in Table A.2, of hydraulic seals and the corresponding test methods. It is to be noted that some of these properties affect the dynamic sealing functions only, some affect the static sealing functions only, and others affect both functions. Additionally, some of the test methods are standardized, and others are based on manufacturer R&D activities but not standardized.

If uniform and repeatable results are to be obtained, it is important to follow the test procedures carefully along with the following considerations:

Test Specimens
Test methods include descriptions of standard specimens for each test. Often, two or more specimens are required due to specimen variability. Results from different specimens are rarely 100% agree.

Test Variables
Test results are only comparable if the following test variables are identical:
- Temperature under which the test was performed.
- Load or pressure used in the test.
- Fluid medium in which the seal is in contact with during the test.
- Duration and rate of applying the test procedure.
- Environmental conditions, such as humidity in the air.

5.1- Resilience

Definition: *Resilience* is essentially the ability of a compound to return quickly to its original shape after a temporary deflection.

Units: It is dimensionless property

Effect: Reasonable resilience is vital to a moving seal. Resilience is primarily an inherent property of the elastomer. It can be improved somewhat by compounding. More important, it can be degraded or even destroyed by poor compounding techniques.

Testing: It is very difficult to create a laboratory test which properly relates this property to seal performance. Therefore, compounding experience and functional testing under actual service conditions are used to insure adequate resilience. . Tested specimens are inspected visually after test.

Hydraulic Sealing Elements and Test Methods
Change in Seal Shape
1- Resilience Test Method: Non-Standard Test
Change in Seal Length
2- Modulus of Elasticity 3- Elongation 4- Yield Tensile Strength 5- Ultimate and Fracture Tensile Strengths Test Method: Standard Test Method (ASTM D412 / DIN 53504)
Change in the Seal Cut
6- Tear Strength: Test Method: Standard Test Method (ISO 34-1 / DIN 53507)
Change in Seal Thickness
7- Compression Set Test Method: Standard Test Method (ASTM D395 / DIN ISO 815)
Change in Seal Volume
8- Swelling Test Method: Standard Test Method (DIN ISO 1817) 9- Shrinkage Test Method: Standard Test Method (TR-10 Low Temperature ASTM D1329-16)
Change in Seal Surface
10- Surface Hardness Test Method: Standard Test Method (ASTM D 2240 / ISO 868 / DIN 53505)
Change in Seal Chemical Structure
11- Compatibility with Hydraulic Fluid Test Method: Standard Test Methods (ASTM D6546-15 OR ISO 6072)
Change in Seal Performance
12-Exrusion Resistance Test Method: Standard Test Method (ASTM C1183 / C1183M) 13-Expolosive Decompression Resistance Test Method: Standard Test Method (NACE TMO192-98) 14- Seal Friction Test Method: Non-Standard Test 15- Wiper Performance Test Method: Non-Standard Test

Table A.2 - Hydraulic Sealing Elements Properties and Test Methods

5.2- Modulus of Elasticity

Definition: As shown in Fig. A.49, *Modulus of Elasticity* of a hydraulic seal is the ratio between the stress (force per cross section area) acting on the seal body and the corresponding percentage longitudinal change of the seal. Also referred to as Young's modulus.

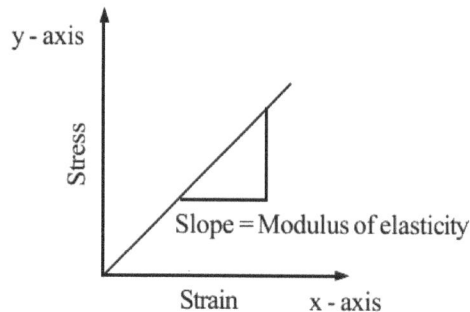

Fig. A.49 – Modulus of Elasticity

Units: It is expressed in Newton per square meter (Pascals Mega Pascals) or pounds per square inch (psi) for a predetermined elongation, usually 100%.

Effect and Importance:
- The higher the modulus of a compound, the more ability to recover after releasing the stress, and the better its resistance to extrusion.
- Modulus of elasticity is directly proportional to the seal hardness.
- Polyurethane and filled PTFE compounds generally have very high tensile strength, providing the associated excellent tear and abrasion resistance.

5.3- Elongation

Definition: *Elongation* is defined as the percentage increase in length with respect to the original length.

Units: It is expressed as a % of the original length. For example, if a 1-inch sample was stretched to two inches, it would be 100% elongation.

Effect and Importance:
- This property primarily determines the stretch which can be tolerated during installation of a sealing element. However, easy to stretch also means easy to extrude.
- The change in the length of a compound is a clear sign of material degradation.

5.4- Yield Tensile Strength

Definitions: *Yield Tensile Strength* is the stress within which the seal maintain elastic performance.

Units: It is expressed in Newton per square meter (Pascals Mega Pascals) or pounds per square inch (psi)

Effect and Importance:
This property determines the maximum strength beyond which the seal is plastically deformed. That helps to know the maximum loads such seal can withstand.

5.5- Ultimate and Fracture Tensile Strength

Definition:
- *Ultimate Tensile Strength* is the stress required to reach ultimate elongation, and seal nicking starts.
- *Fracture Tensile Strength* is the stress at which the seal breaks.

Units: It is expressed in Newton per square meter (Pascals Mega Pascals) or pounds per square inch (psi)

Effect and Importance:
Elongation and all Tensile Strength values are used throughout the industry as a quality assurance measure on production batches of elastomer materials.

Testing: (Standard Test Method ASTM D412 / DIN 53504)
Test Purpose: Determination of Tensile Strength values.

Test Specimen: The test specimens used for this purpose are usually tensile bars or standardized rings with rectangular specific cross-sections. Inconsistent values were found as a result of using specimens with various cross sections.

Test Sequence: As shown in Fig. A.50, the test specimen is stretched at a constant speed to break. Tensile strength curve (force versus change in the length) is developed.

In pulling specimens to find tensile strength, elongation, and modulus values, ASTM D412 requires a uniform rate of pull of 508 mm (20 inches) per minute. In one test, tensile strength was found to decrease 5% when the speed was reduced to 50.8 mm (2 inches) per minute, and it decreased 30% when the speed was further reduced to 5.08 mm (0.2 inches) per minute. Elongation and modulus values decreased also, but by smaller amounts.

High humidity in the air will reduce the tensile strength of some compounds.

<u>Test Evaluation:</u> As shown in Fig. A.51, the following values can be calculated from the experiment:

- Yield Tensile Strength.
- Ultimate Tensile Strength.
- Fracture Tensile Strength.
- Elongation at break.

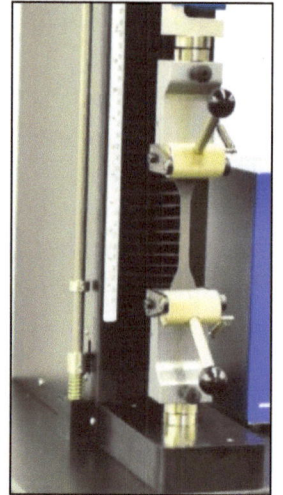

Fig. A.50 - Elongation and Tensile Strength Test Machine

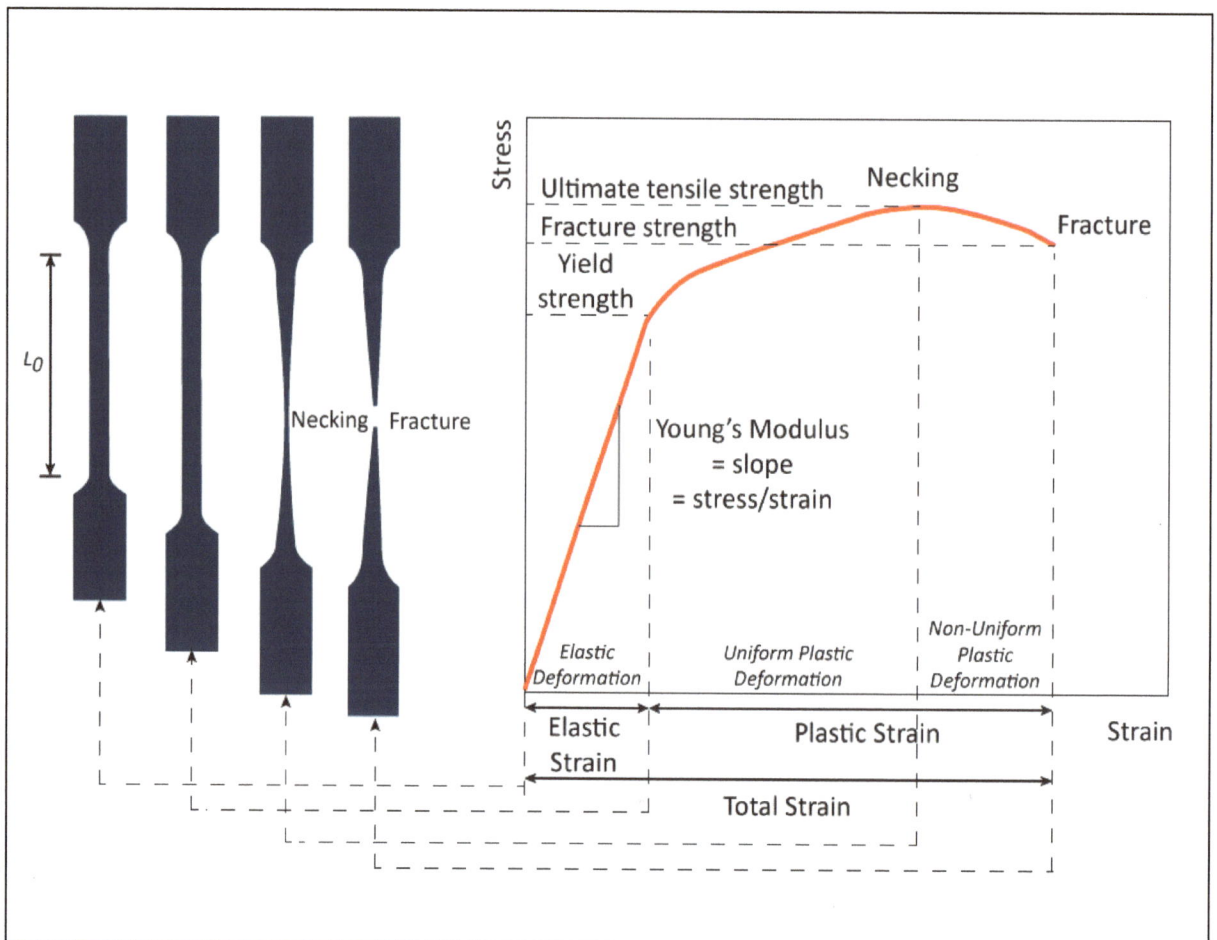

Fig. A.51 - Results from Elongation and Tensile Strength Test

5.6 -Tear Strength

Definition: *Tear Strength* is the ratio of the force achieved at the moment of rupture and the initial cross-section thickness of the specimen.

Units: Tear Strengths is expressed in pounds/inch (1b/in) or Kilo Newton/meter (kN/m).

Effect and Importance:
Tear Strength is important for dynamic seals and need not be considered for static applications. Seals with poor tear resistance (less than 100 lbs./in.) (17.5 kN/m):

- Have the danger of tearing during assembly if it must pass over ports and sharp edges.
- Will fail quickly under further flexing or stress once a crack is started.

Testing (Standard Test Method ISO 34-1 / DIN 53507):
Test Purpose: Determination of the *Tear Propagation Resistance* (the sensitivity of elastomers in the event of cutting and tearing damage).

Test Specimen: As shown in Fig. A.52, the specimens used for this test are usually tensile bars or standardized rings with rectangular cross-sections.

Test Sequence: A longitudinal cut is made in the material to be tested, the two half-strips are clamped in a pulling machine and pulled apart.

Test Evaluation: The force required to propagate the cut is measured in relation to the sample thickness.

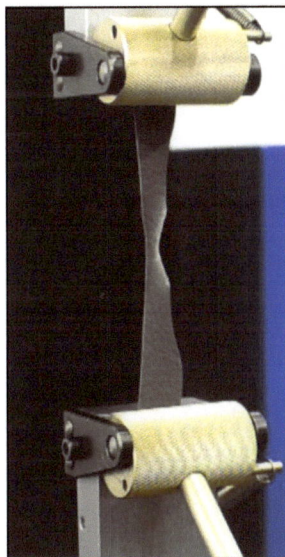

Fig. A.52 - Tear Resistance Test

5.7- Compression Set

Definition: The *Compression Set* (CS) refers to the ability of a sealing material to restore thickness and consequently the sealing force after a certain time in contact with a fluid medium under certain temperature.

Units: It is expressed as a % in comparison with a targeted value of deformation.

Effect and Importance:
The lower the compression set value, the better restoring, sealing, and seal lifetime. In general, Compression Set is a result of one or more of the following conditions:
- Selection of seal material with inherently poor compression set properties.
- Improper gland design.
- Excessive temperature causing the seal to harden and lose its elastic properties.
- Volume swell of the seal due to system fluid.
- Excessive seal squeeze due to over tightening of adjustable glands.
- Incomplete curing (vulcanization) of seal material during production.
- Fluid incompatibility with the seal material.

Note: Magnitude of CS is inversely proportional to the ability or recovery. Hence, Poor CS means higher magnitude. Figure A.53 shows a flattened O-Ring due to poor compression set. The seal shown in Fig. A.54 exhibits nearly 100% compression set, i.e. no recovery (thickness restoration) after compression.

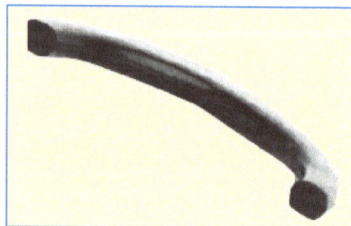

Fig. A.53 - Flattened O-Ring due to Significantly Poor Compression Set (Courtesy of Parker)

Fig. A.54 – Deformed Seal due to Poor Compression Set (Courtesy of Parker)

Testing (Standard Test Method ASTM D395 / DIN ISO 815):

Test Purpose: Determination of the compression set

Test Specimen: a specimen with a specific form.

Test Sequence (As shown in Fig. A.55):
- Compression test must not be done earlier than 16 hours after elastomer manufacturing.
- Tests should only be carried out with samples which have not been previously stressed.
- Original thickness (h_0) of the test specimen is measured.
- The sample is compressed to the thickness (h_1) (default: $h_1 = 0.75\ h_0$).
- The compressed specimen is stored in an apparatus in a medium (default: air) and under a specified temperature for a predefined test time (default: 24 or 72 hours).
- Test temperature should be mentioned in the test report.
- 30 minutes after removal from storage and removing the compression, recovered thickness of the specimen (h_S) is measured.

Test Evaluation:
Compression set values are only comparable if the following test parameters are identical:
- Form of the test specimen.
- Deformation (default: 25%).
- Duration of the deformation.
- Temperature and medium during the deformation.
- Equation A.2 shows how to calculate the compression set value.
- ASTM Compression Set D395 Test Method B, states, "The percentage of compression employed shall be approximately 25%." Significantly higher compression set values were found after compressing less than 25%, while results after 30 or 40% compression were sometimes smaller.

$$\mathbf{CS}\ (\%) = \frac{(\mathbf{h_0 - h_S})}{(\mathbf{h_0 - h_1})} \times \mathbf{100} \qquad\qquad \mathbf{A.\ 2}$$

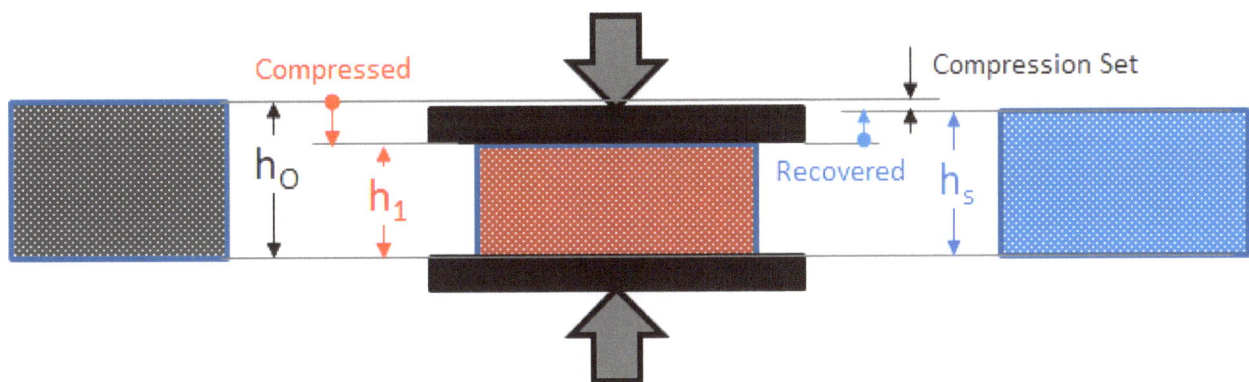

Fig. A.55 - Compression Set Test

5.8- Swelling

Definition: Elastomers have a higher coefficient of thermal expansion than steel. This means the seal will expand more when hot. *Swelling* of a hydraulic seal is the volumetric increase of the seal after it has been in contact with a fluid medium.

Units: It is expressed as a % of the original volume.

Effect and Importance: Seal Swelling is associated with changes in the seal physical properties such as:
- Possibility of seal extrusion under high pressure.
- Reduced hardness, elasticity, and tensile strength.
- Marked softening of the elastomer.

The magnitude of the volume change depends on five factors:
1. Fluid in contact
2. Composition of the elastomeric compound.
3. Working conditions (temperature, pressure, humidity, and time).
4. Geometric form (thickness) of the seal.
5. Stress condition of the seal. (Volume change of stretched parts is greater than compressed parts).

Testing (Standard Test Method DIN ISO 1817):
Test Purpose: Determination of seal volume increase after a defined storage period in contact with specific fluid under certain temperature.

Test Sequence:
1. The volume of the seal is determined.
2. The seal is immersed in a test fluid and stored according to the standard or to customer specifications.
3. At the end of the storage period (and after cooling down), the volume of the seal is measured again.
4. The result is expressed as a percentage of the initial volume.

Test Evaluation: As a rule-of-thumb (unless otherwise stated).
- For static seals, up to 50% volume swell can usually be tolerated.
- For dynamic applications, (10-20) % swell is a reasonable range.
- Seals with smaller cross-sections have been found to swell more than larger ones.

5.9- Shrinkage

Definition: Elastomers have a higher coefficient of thermal expansion than steel. This means the seal will shrink more when cold than steel does. *Shrinkage* of a hydraulic seal is the volumetric decrease of the seal after it has been in contact with a fluid medium.

Units: It is expressed as a % of the original volume.

Effect and Importance: Seal Shrinkage is associated with changes in the seal physical properties such as:
- As in seal swelling, reduced hardness, elasticity, and tensile strength.
- As in seal swelling, marked softening of the elastomer.
- Reduced retaining force between the seal and the static housing faces.

Testing (Low-Temperature Standard Test Method ASTM D1329-16 / TR-10):
Test Purpose: Determination of elastomer retraction and viscoelastic properties at low temperature.

Test Specimen:
- Figure 4.56 shows several specimens of various lengths mounted and ready to be elongated. Standard specimens have gage lengths of 1", 1.5", or 2".

Test Sequence:
- A small dog-boned specimen is held in an elongated condition.
- The test fixture is used to apply 100% elongation.
- The specimen is subjected to a low temperature for certain time.
- The specimen is then allowed to retract freely while raising the temperature at a uniform rate of 1°C (1.8 °F) per minute.
- Measure the temperature at 10% and 70% retraction (shrinkage).
- Comparison is based on the % shrinkage at a specific temperature or the temperature at specific % shrinkage.

Test Evaluation: Elastomers are compared based on temperatures corresponding to 10% and 70% because this is the temperature window within which the elastomers start and end the retraction. Maximum tolerated shrinkage for both static and dynamic seals is (3-4) %.

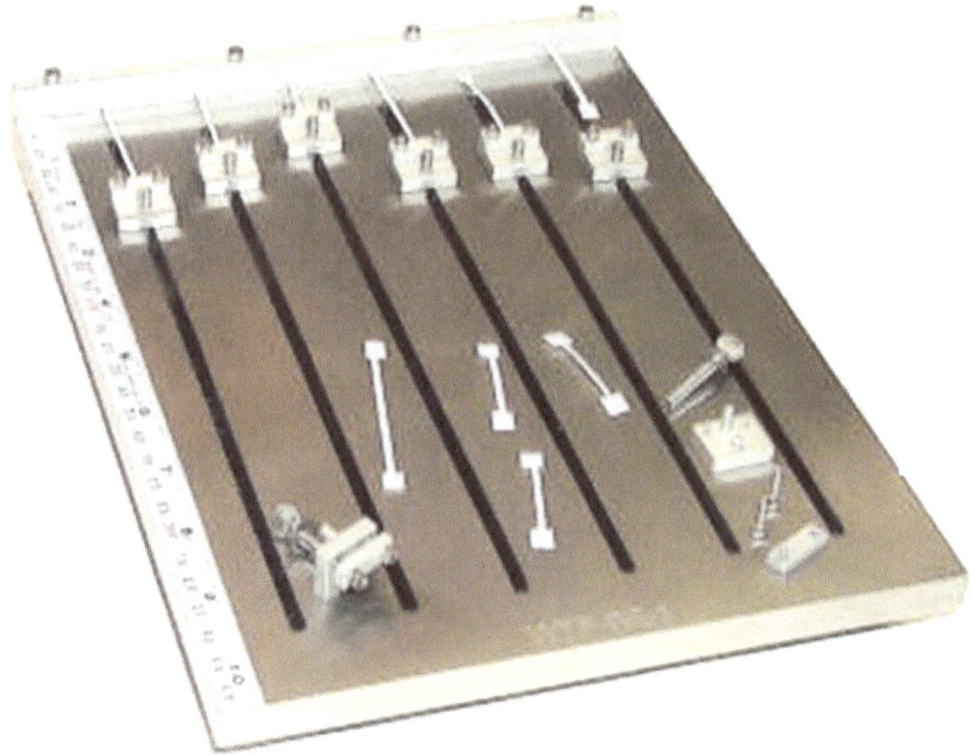

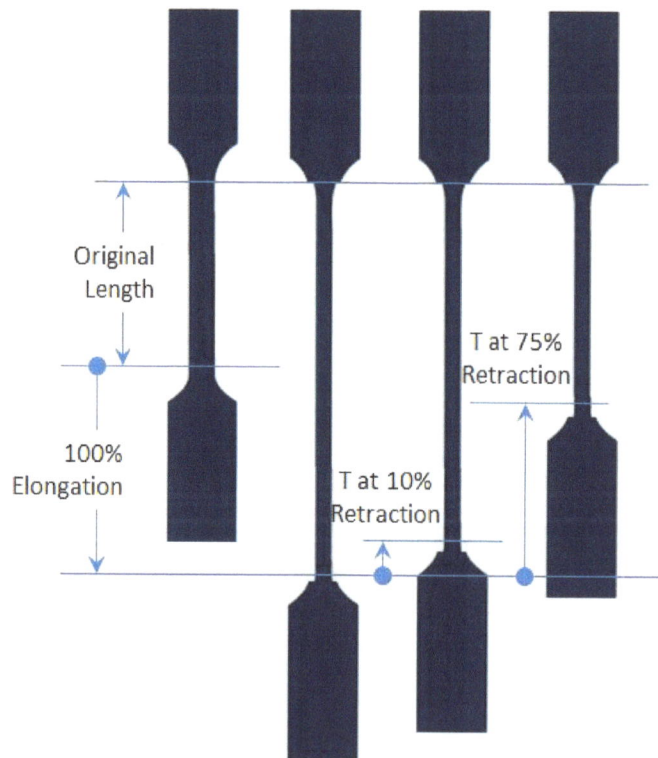

Fig. A.56 - Elastomers Retraction Test (www.wyomingtestfixtures.com)

5.10- Surface Hardness

Definition: One of the most important parameters in rubber technology is the *Hardness*. Hardness, also referred to as *Durometer*, is the resistance of a body against penetration of a harder body of a standard shape at a defined load.

Units: The hardness scale has a range of 0 (softest) to 100 (hardest). Common hydraulic seals have hardness that fall in two scales, *Shore A* (for soft-to-medium compounds) and Shore D for (for medium-to-hard compounds).

Effect and Importance:
This property is important to dynamic seals where harder elastomers have better ability to resist abrasion, wear, and surface scraping. For static seals, higher hardness helps avoid seal extrusion. It is also to be noted that:
- Seal hardness is directly proportional to the Modulus of elasticity.
- Seal leakage is directly proportional to seal hardness.
- Seal friction is inversely proportional to seal hardness.

Testing:
- For Shore A" and "Shore D": **Standard Test Methods ASTM D 2240 / ISO 868 / DIN 53505.**
- For "Durometer IRHD (International Rubber Hardness Degrees): **Standard Test Methods ASTM 1414 and 1415 / ISO 48**.

Test Purpose: determination of seal hardness

Test Specimen:
- Diameter min. 30 mm (1.181 inch).
- Specimens of softer elastomers should be thicker than harder Elastomers. Thickness min. 6 mm (0.240 inch).
- Upper and lower surfaces smooth and flat.
- The samples must not have been previously stressed.

Hardness Testers:
As shown in Fig. A.57, the test can be done by manual or automated *Hardness Tester*.

The Shore type hardness testers are spring loaded indentation devices, in which values are obtained as a function of the viscoelastic property of the material. A calibrated spring force is applied on a standard pin (Indentor) against the specimen. Each 0.001 inch (0.0254 mm) of deflection of the pin is shown as 1-degree Shore (A). Therefore, harder material results in more material resistance to penetration, more spring compression and pin deflection, which means higher shore number.

As shown in Fig. A.58, the shape of the pins depends on the estimated range of hardness. For Shore A, pin with conical head. For Shore D, pin with spike head.

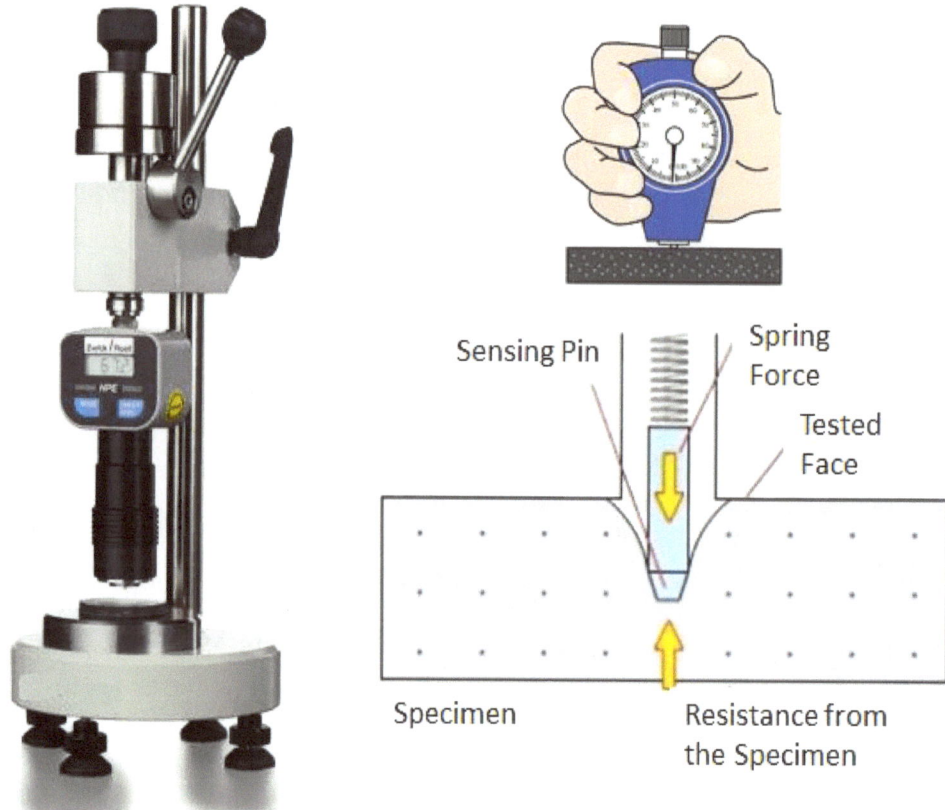

Fig. A.57 - Hardness Test Devices for Elastomers

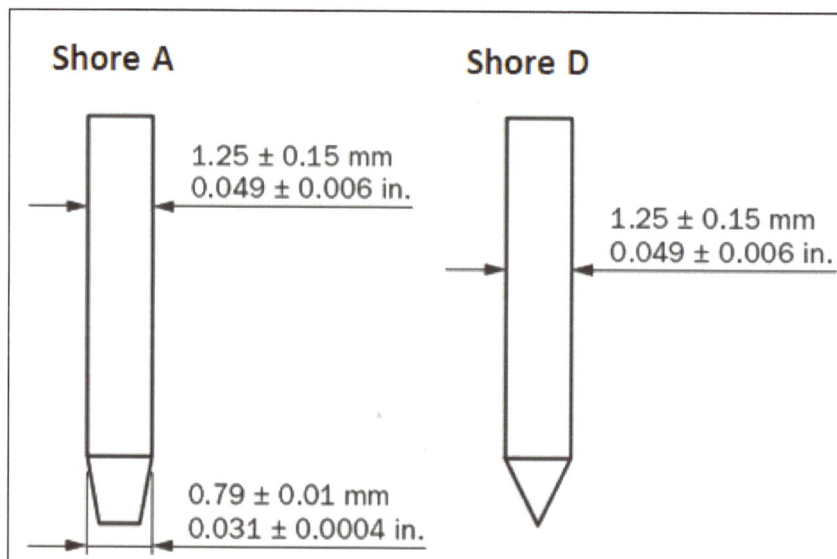

Fig. A.58 - Pins for Hardness Tests (Courtesy of Trelleborg)

Test Sequence:
- The test must not be carried out earlier than 16 hours after elastomer manufacturing.
- The test should be carried out at temperatures of 23 ±2 °C (73.4 ±2 °F).
- The test temperature should be mentioned in the test report.
- The test must not use samples that have been previously stressed.
- In gauging the hardness of an O-Ring, which has no flat surface, accuracy of the test may vary depending on the measuring spot. As shown in Fig. A.59, the actual crown of the O-Ring, the point that gives the most reliable reading.
- The test is carried out by applying a specific load on the pin at a certain speed. Inconsistent results were found as a result of performing the test at different speeds and different loads.
- The value is read after a holding time of three seconds.

Test Evaluation (as shown in Table A.3):
- Shore A values of 60-75 are recommended for O-rings.
- Shore A values of 75-90 are recommended for seals and are more resistant to abrasion.
- Harder compounds (above Shore A 90), Shore D measurements must be made.
- The scales overlapped in between the medium and hard rubbers. For example, Shore A 90 approximately equal Shore D 40.

Fig. A.59 - Hardness Test for O-Rings

Shore A														
5	10	20	30	40	50	60	70	80	90					
Shore D														
									40	50	60	70	80	85
Very Soft			**Soft**			**Medium**				**Hard**				

Table A.3 - Hardness Scales for Hydraulic Seals

5.11- Compatibility with Hydraulic Fluids

Definition: *Compatibility* with hydraulic fluids means the ability of a seal to resist chemical interaction with the working hydraulic fluids.

Effect and Importance:
This property is very important because any change in the seals chemical structure significantly affects the seal shape and physical properties. Seal deterioration due to chemical reaction with the hydraulic fluid causes clogging of control orifices, leakage, and the relevant consequences.

Table A.4 shows the state of compatibility of common seals with common hydraulic fluids.

Seal materials	Fluid Types					
	Petroleum oil	Water-in-Oil Emulsion	Water Glycol	Phosphate Ester*	Chlorinated hydrocarbon	Synthetic with petroleum fractions
Buna-N (Acrylonitrile)	Excellent	Excellent	Very Good	Poor	Poor	Poor
Neoprene (Chloroprene)	Good	Good	Good	Poor	Poor	Poor
Butyl	Poor	Poor	Good	Fair to good	Poor	Poor
Silicone	Fair	Fair	Fair to poor	Fair to good	Poor to fair	Fair
Ethelene-Propylene	Poor	Poor	Good to excellent	Excellent	Fair	Poor
Viton® (Fluorocarbon)	Excellent	Excellent	Excellent	Good to Excellent	Good to Excellent	Good to Excellent
Metals	Conventional	Conventional	**	Conventional	Conventional	Conventional
Pipe Sealants	Conventional, Loctite® or Teflon® tape	Conventional, Loctite® or Teflon® tape	Loctite® or Teflon® tape	Loctite® or Teflon® tape	Loctite® or Teflon® tape	Loctite® or Teflon® tape

- *Many types and blends of fluids are sold under the designation "phosphate ester." Check with fluid supplier to verify exact compatibility.
- **Avoid zinc, cadmium, or galvanized materials.
- Viton® and Teflon® are trademarks of E.I DuPont DeNemours & Co., Inc.
- Loctite® is a trademark of the Loctite Corp.

**Table A.4 - Compatibility of Common Hydraulic Fluids with Common Seal Materials
(www.schoolcraftpublishing.com)**

Testing (Standard Test Methods ASTM D6546-15 and ISO 6072):

Test Purpose: This test is used for determining compatibility of elastomeric seals for industrial hydraulic fluid applications.

Test Procedure: The test procedure, as shown in Fig. A.60, includes exposing an O-ring test specimen to industrial hydraulic fluids under definite conditions of temperature and time.

Test Evaluation: The resulting deterioration of the O-ring material is determined by comparing the changes in work function, hardness, physical properties, compression set, and seal volume after immersion in the test fluid to the pre-immersion values.

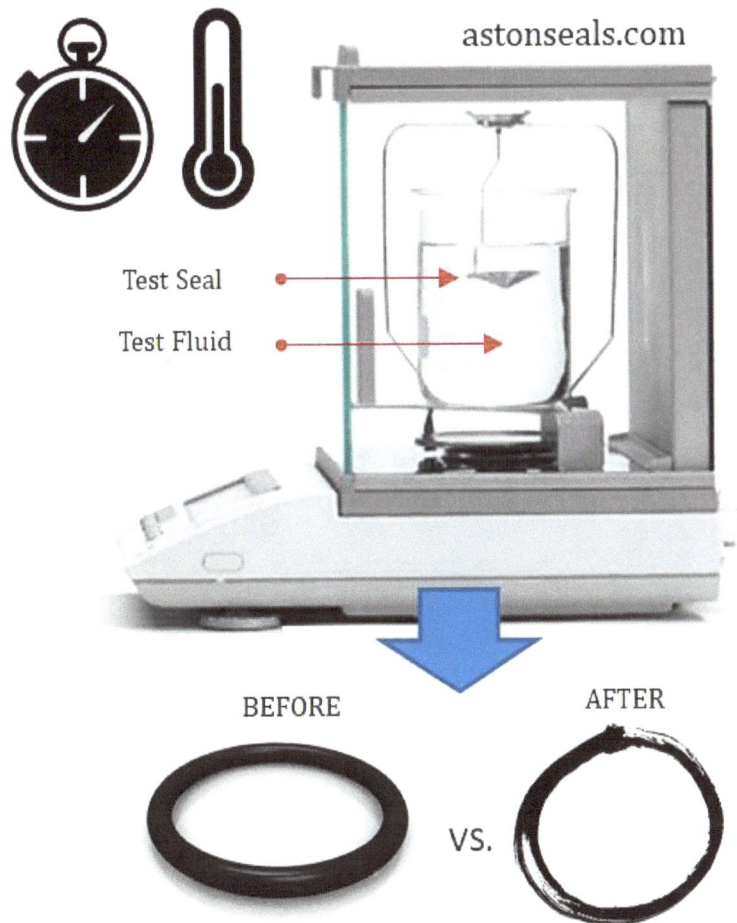

- Changes in work function.
- Hardness.
- Physical properties.
- Seal volume.

Fig. A.60 - Fluid Compatibility Standard Test Method

5.12- Extrusion Resistance

Definition: *Extrusion Resistance* is the seal's ability to resist extrusion through the gap between sealed surfaces as a result of increased working pressure or seal material softness.

Effect and Importance:
Seal extrusion causes seal failure, excessive leakage, and oil contamination.

Testing: (Standard Test Method ASTM C1183):
Test Purpose: This test method covers the procedures for determining the extrusion rate of elastomeric seals. There is no known ISO equivalent to this test method.

Test Procedure: This test method measures the volume of the seal extruded over a given period of time at a given pressure (kPa or psi).

5.13- Explosive Decompression Resistance

Definition: *Explosive Decompression Resistance* is the seal's ability to resist damage due to gas bubbles rapid decompression from a high pressure.

Effect and Importance:
'Explosive Decompression' is a commonly used term for the damage caused when a rubber seal, containing absorbed gas, is subjected to rapid decompression from a high pressure. The failure of a seal due to explosive decompression damage can lead to obvious safety and lost production consequences.

Testing: (Standard Test Method NACE TMO192-98):
Various oil companies and national standards organizations have developed test protocols aimed at defining elastomeric seal materials that can avoid damage during decompression in high pressure gas duty. This test is used for evaluating elastomeric materials in Carbon Dioxide decompression environments. Finite element analysis potentially provides a tool for reducing the quantity of costly and time-consuming performance testing which currently has to be carried out to prove seal integrity.

5.14- Hydraulic Seal Friction

5.14.1- Hydraulic Seal Friction Conditions

As shown in Fig. A.61, hydraulic seal friction depends on the lubrication condition between the seal and the sealed surface as follows:

- Boundary seal friction if no lubricant is found between the seal and the sealed surface.
- Mixed friction if little lubricant is found between the seal and the sealed surface.
- Viscous friction if the lubricating film is able to completely separate the seal from the sealed surface.

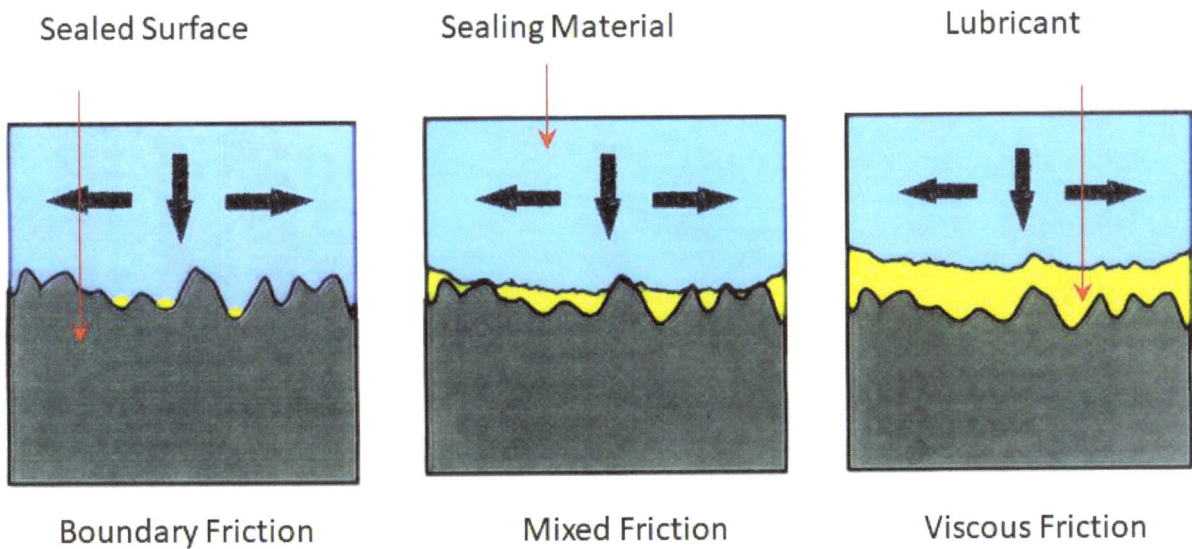

Fig. A.61 - Hydraulic Seal Friction Conditions

5.14.2- Friction in Translational Seals

Examples of translational seals are cylinder rod and piston seals. This friction is usually mistakenly ignored in calculating a cylinder force or pressure. For a cylinder speed ranging from 5-15 cm/s (10 – 30 fpm), overcoming cylinder seal friction wastes approximately 5% of the cylinder input power. Seal leakage in hydraulic cylinders is approximately one tenth of the losses due to friction.

Figure A.62 shows results of experimental work to investigate extending rod seal friction versus velocity at 75 bar (1088 psi) constant pressure. The figure shows that various seals have the same trend. Seal friction drops once the rod starts extending, then gradually increases with increasing velocity.

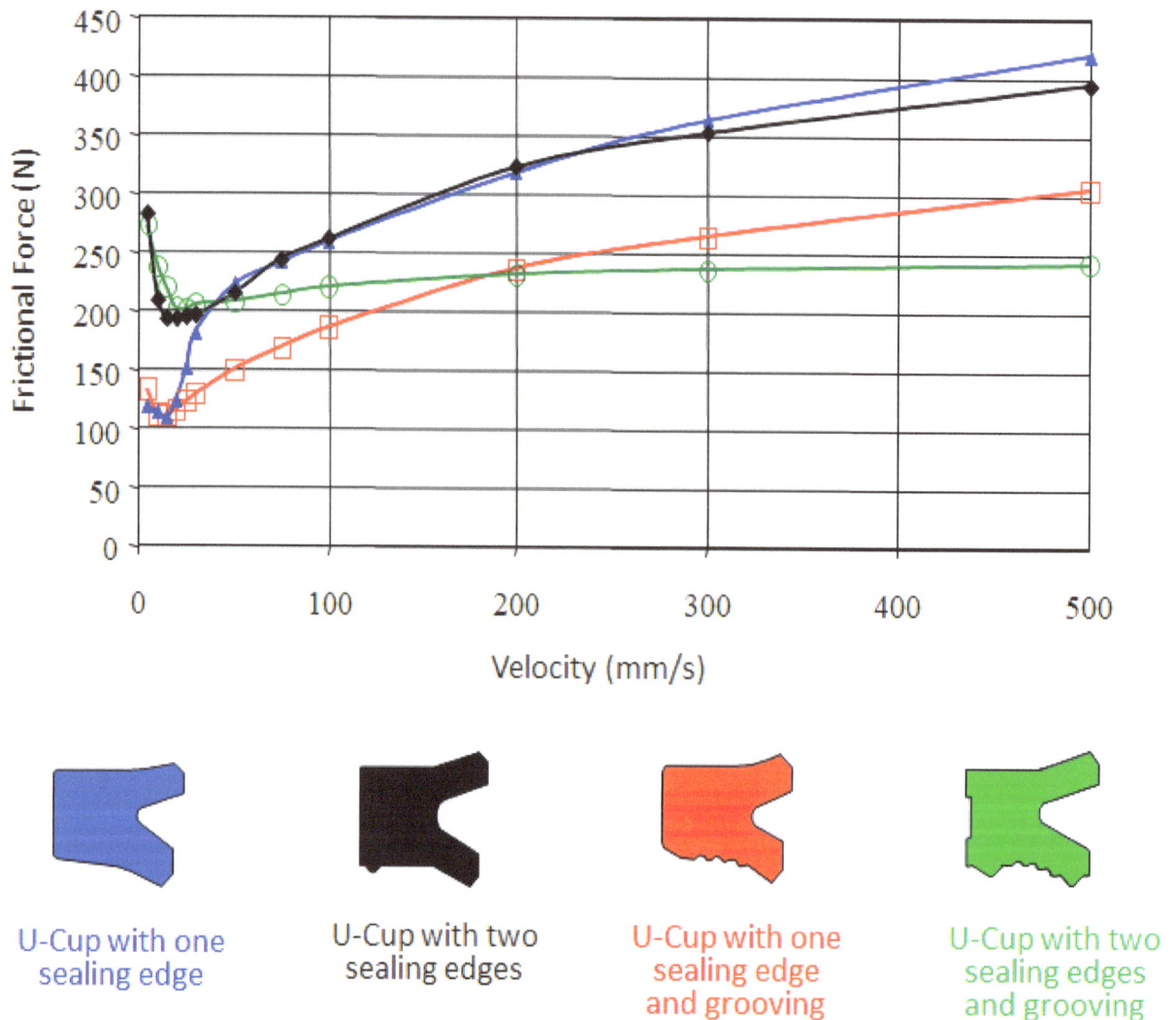

U-Cup with one sealing edge

U-Cup with two sealing edges

U-Cup with one sealing edge and grooving

U-Cup with two sealing edges and grooving

Fig. A.62 - Rod Seal Friction at Constant Pressure

5.14.3- Friction in Rotational Seals

Examples of rotational seals are in hydraulic pumps and motors. Figure A.63 and Equation A.3 show how to calculate the surface (*Tangential*) speed on the sealed surface of a rotational shaft.

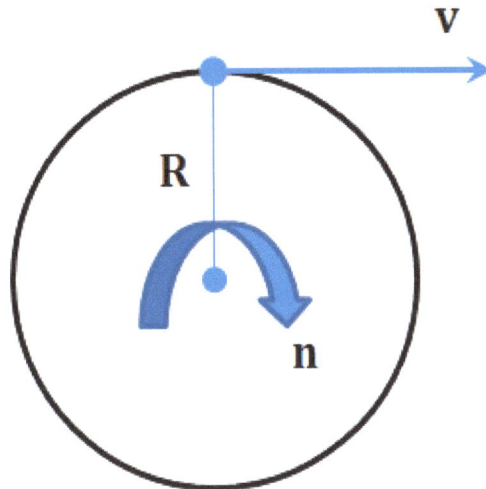

Fig. A.63- Surface (Tangential) Speed of a Rotational Shaft

$$\mathbf{v} = \omega \times R = 2\pi n \times R \hspace{3cm} \text{A.3}$$

Where:

$\mathbf{v}$ = Tangential speed in (m/min) or (FPM) depends on the units of $\mathbf{R}$.
$\mathbf{R}$ = Radius of the shaft in (m) or (foot).
$\mathbf{n}$ = Shaft rotational speed (RPM).

Figure A.64 shows a Nomograph that was developed to provide a quick and approximate solution for the previous equation. For example:
- If shaft rotational speed = 4000 rpm, and shaft Diameter = 1 (in) ≈ 25 (mm), then:
- Surface speed (FPM) = 2 x 3.14 x 4000 x 0.5 / 12 = 1047
- Surface speed (m/min) = 2 x 3.14 x 4000 x 0.025 / 2 = 314

As shown in the previous example, surface speed is 314 m/min = 523 cm/s. this speed is almost 100 times greater than the recommended linear speed of a cylinder rod. Therefore, seal friction in rotational shafts is much higher than in translational shafts.

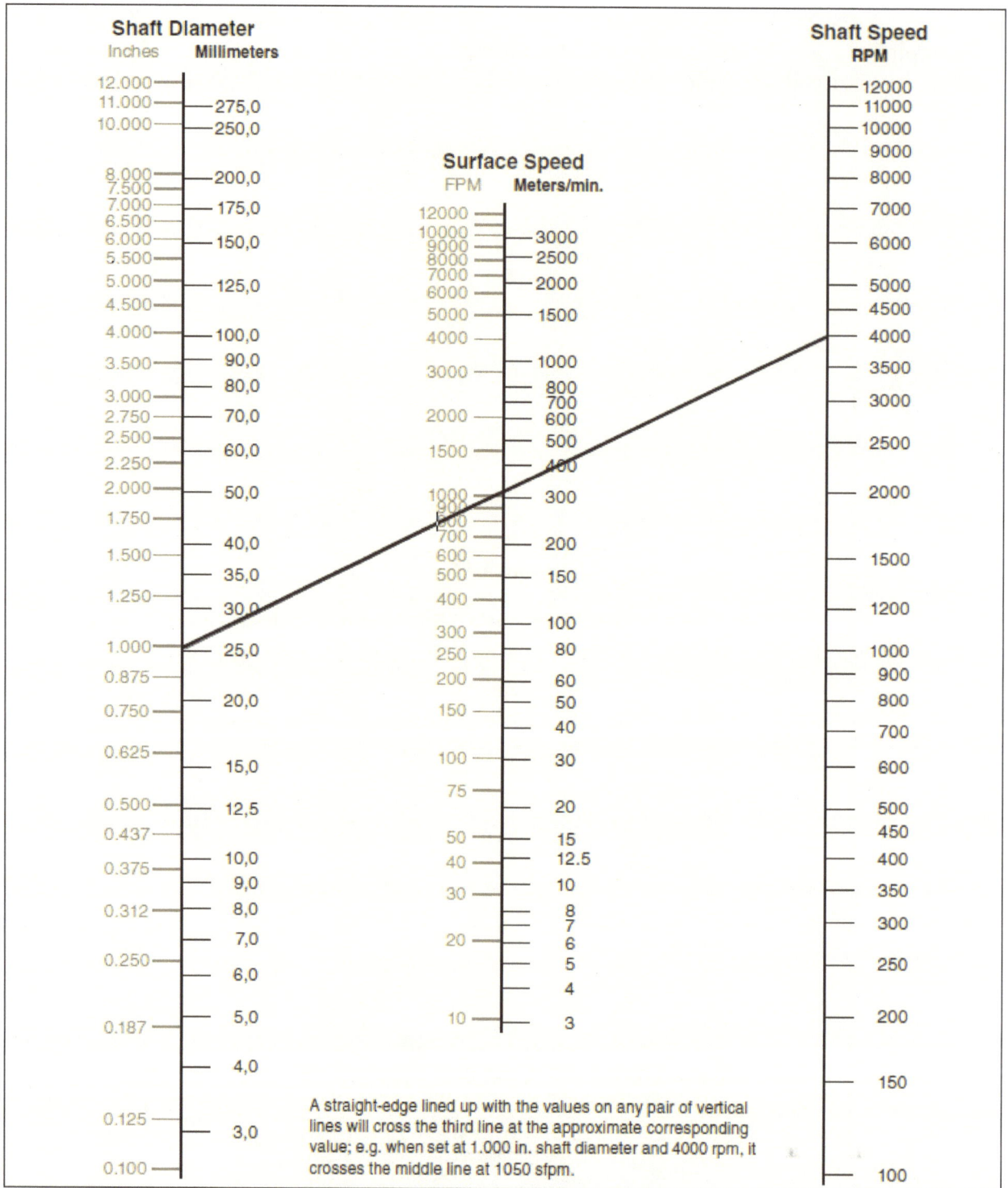

Fig. A.64 - Nomograph for Calculating Rotational Shafts Surface Speed (Courtesy of Parker)

5.14.4- Factors Affecting Seal Friction

Friction of dynamic seals depends primarily on:
- **Seal Geometry:** shape, dimensions, exposed area, and dynamic surface roughness.
- **Seal Material:** type and properties.
- **Working Conditions:** temperature and pressure.
- **Speed (translational or rotational):**
  - **Low Speed:** affects the performance of the seal.
  - **High Speed:** causes a breakdown in the oil film between the seal and the sealing surface, and, therefore, the seal runs dry which leads to premature seal failure.

5.14.5- Controlling Seal Friction

- ❖ Basically, seal friction can't be 100% eliminated. However, lowering seal friction results in:
  - Reduced energy loss and heat generation.
  - Reduced wear.
  - Reduced rate of chemical attack.
  - Increased extrusion resistance.
  - Increased seal life.

- ❖ Seal friction can be reduced by a combination of the following:
  - Seal Design:
  - Waved seal surfaces to retain lubricating oil.
  - Seal design and placement should consider better heat dissipation.

  - Seal Material:
  - Seals that work at low speed are coated by a dry lubricant such as Teflon layer.
  - Seals with high hardness have less fiction. However, seal hardness should be compromised with seal leakage.

  - Working Conditions:
  - Select proper seals based on field working conditions.
  - Control working conditions within recommended limits.
  - Use compatible fluids with anti-friction additive packages.

5.15- Wiper Performance Test

Effect and Importance:
Dirt ingression is a major cause of hydraulic system inefficiency, degradation and failure. Although superior filtration systems exist and are used to limit exposure to contaminants, there are several locations in a typical hydraulic system that remain vulnerable. Breather ports, external couplers and power shaft systems that drive hydraulic pumps, external motors and cylinders are included among areas that may be compromised.

Testing:
Test Purpose: A new innovative repeatable test method, partially based on SAE J1195, has been developed by **Hallite Seals Americas**, Inc. in cooperation with **Milwaukee School of Engineering** (MSOE) to assess the amount of dirt entering a simulated hydraulic system through the rod wiper located on the hydraulic cylinder.

Test Conditions:
MSOE's Fluid Power Institute™ built a test rig with the following parameters:
  ▪ Test Duration: 24.000 cycles, 24384 meters (80,000 feet) linear travel.
  ▪ Cycle Rate: 0.25 Hz
  ▪ Total Stroke Length: 101.6 cm (40 inches).
  ▪ Test Pressure: Atmospheric
  ▪ Test Temperature: 66 °C (150 °F).
  ▪ Test Oil: MIL-PRF-46170
  ▪ Test Contaminant: ISO 12103-1-A4 Course Test Dust

Test Procedure (as shown in Fig. A.65):
  ▪ The test rod wiper was installed in the rod end of the cylinder housing along with a TPE-faced, two-piece rod seal (Hallite Type 16 profile) to simulate typical boundary lubrication that is found on the rod in standard cylinder application.
  ▪ The hydraulic fluid was heated to 150°F by heaters located in the reservoir.
  ▪ The hydraulic fluid is pumped beneath the rod wiper.
  ▪ Air supplied to the dust chamber. Dust moves at a high velocity inside the chamber, as air is forced in
  ▪ The rod is cycled with the specified rate for the specified duration.
  ▪ Oil drains back to reservoir.
  ▪ Dirt content measured in the oil reservoir via the particle counter.

Test Evaluation:
MSOE tested the Hallite 520 and 820 wipers against two competitors. Based on these results, the Hallite 820 provided the most protection allowing the least amount of contamination into the test fixture. Note, the Hallite 820 utilizes a secondary protective structure outboard of the primary wiper lip (referred to Hallite's Umbrella Wiper Technology) which accounts for the better ingression protection. The Hallite 520 also provided a high level of protection against contamination as compared to similar competitor wipers.

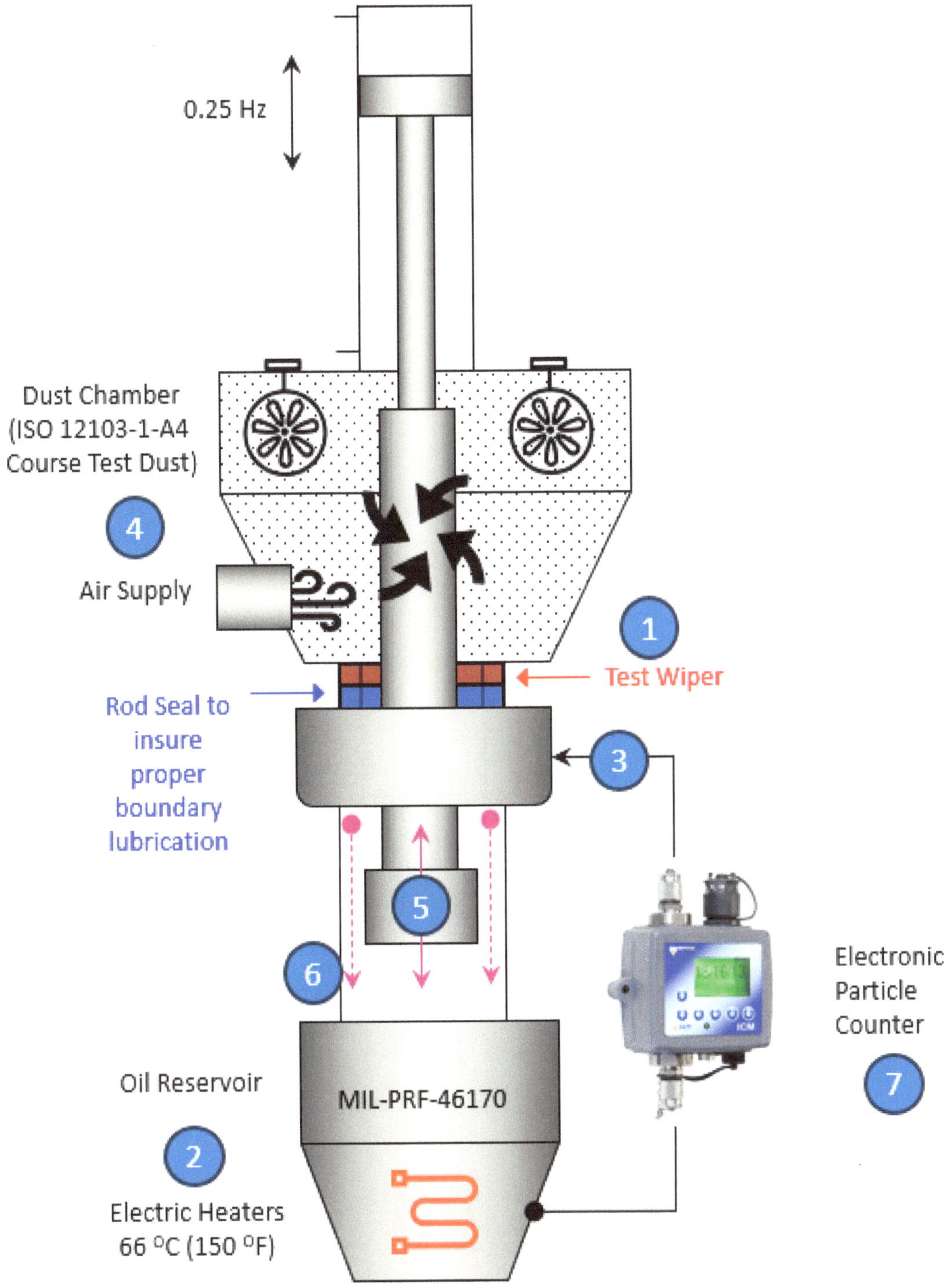

Fig. A.65 - Cylinder Rod Wiper Performance (Courtesy of Hallite Seals)

Chapter 6 – Best Practices for Hydraulic Seals Selection

Hydraulic seals selection is a major step in designing a hydraulic component. The best hydraulic seal is one that:

- Performs efficiently with low friction.
- Works under extended operating conditions (pressure and temperature).
- Has high tensile strength and resists twisting and spiral failures.
- Has high tear and abrasion resistance.
- Compatible with various types of hydraulic fluids.
- Resists chemicals and acids.
- Requires minimum space and is easy to install.
- Cost effective.

Unfortunately, there is no one seal that satisfies all the sealing requirements. Therefore, seal properties should be compromised based on the application. The following best practices list provides guidelines for selecting hydraulic seals.

1. Selection of Seal Type.
2. Selection of Seal Dimensions.
3. Selection of Seal Lip Geometry.
4. Selection of Seal Crossectional Shape.
5. Selection of Seal Material Based on Working Temperature.
6. Selection of Seal Material Based on Working Pressure.
7. Selection of Seal Material Based on Working Fluid.
8. Selection of Seal Material Based on Hardness.
9. Selection of Seal Material Based on General Properties.

The following subtitles provide interpretation for the items contained in the best practice list.

6.1- Selection of Seal Type

Figure A.66 shows a guide for selecting the general type of hydraulic seal. As shown in the figure, the general type of the seal depends on the application and the sealed components.

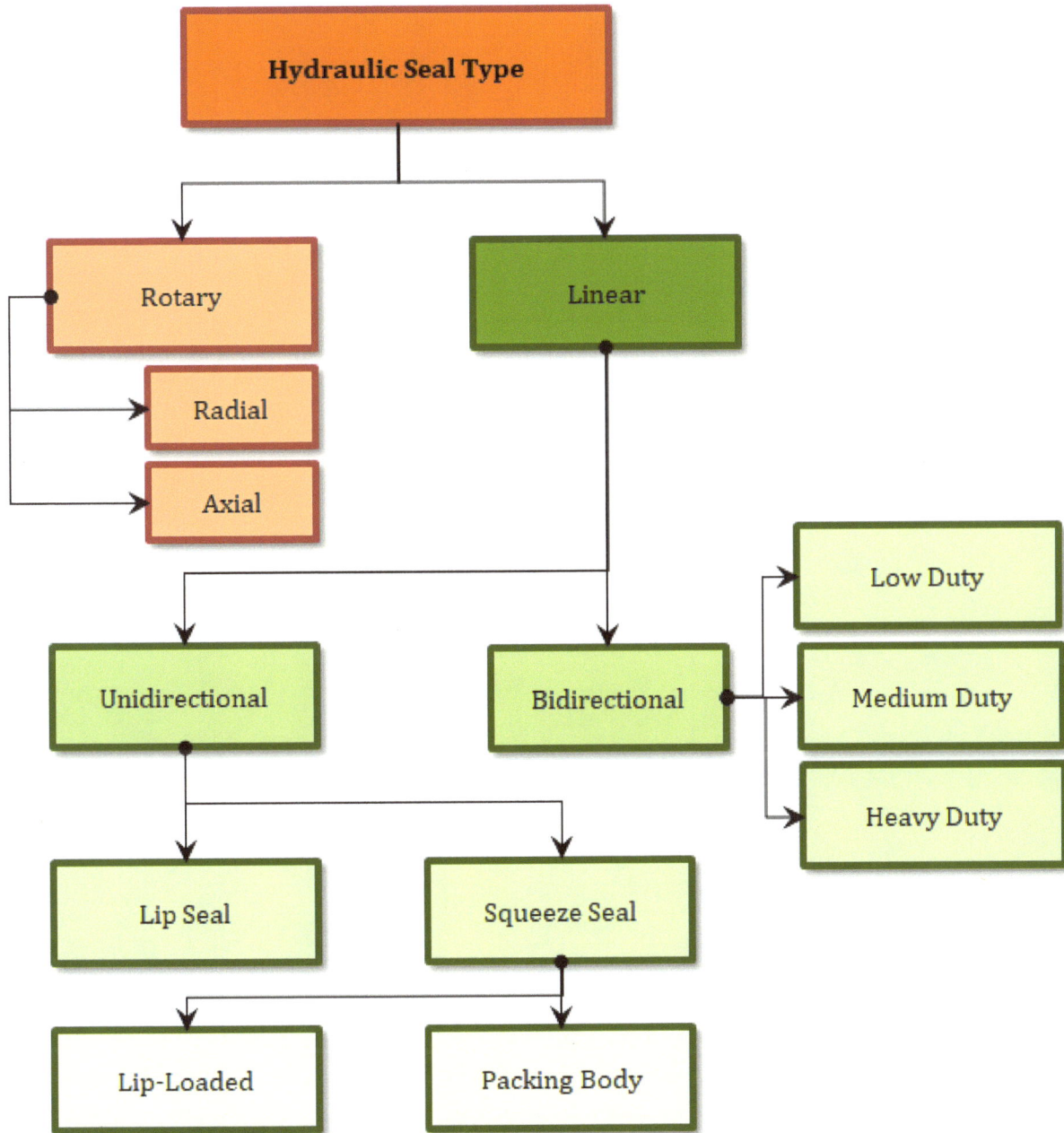

Fig. A.66 - Hydraulic Seal Type Selection

6.2- Selection of Seal Dimensions

O-Ring Cross Section versus its Stability: In designing an O-ring seal, there are usually several standard cross section diameters available. For any given piston or rod diameter, O-rings with smaller cross section diameters are inherently less stable than larger cross sections, tending to twist in the groove when reciprocating motion occurs. This leads to early O-ring spiral failure and leakage.

6.3- Selection of Seal Lip Geometry

Lip geometry will determine several functions of the seal. Force concentration on the shaft, film breaking ability, hydroplaning characteristics and contamination exclusion are all factors dependent on lip shape. Table A.5 shows four different lip shapes and provides helpful insights for choosing an appropriate lip geometry as follows:

- **Example 1:** Rounded Cut is recommended for film stability, and self-lubrication.
- **Example 2:** *Straight Cut* is recommended for best contamination exclusion, Rod Wipers.
- **Example 3:** *Beveled Cut* is recommended for best film breaking for rod sealing.
- **Example 4:** *Square Cut* is recommended for piston guiding and sealing.

Contact Shape	Rounded	Straight Cut	Beveled	Square
Seal Lip Shape Shape of Contact Force/ Stress Profile				
Film Breaking Ability	Low	High	Very High	Medium
Contamin-ation Exclusion	Low	Very High	Low	High
Tendency to Hydroplane	High	Very Low	Low	Medium
Typical Uses	Pneumatic U-cups	Wipers and Piston Seals	Rod Seals	Piston Seals

Table A.5 - Effect of Lip Geometry on Seal Function (Courtesy of Parker)

6.4- Selection of Seal Crossectional Shape

The crossectional shape of a seal dramatically affects the sealability, especially at low pressure. Figure A.67 shows that, for dynamic seals, low friction performance is traded off with the sealability at low pressure. In other words, the seal cross section that offers better sealability at low pressure experiences increased friction.

With this in mind, seals are often categorized as either "*Lip Seals*" or "*Squeeze Seals,*" and many other seals fall somewhere in between. Lip seals are characterized by low friction and low wear; however, they also have poor low pressure sealability. Squeeze seals are characterized by just the opposite: high friction and high wear, but better low pressure sealability.

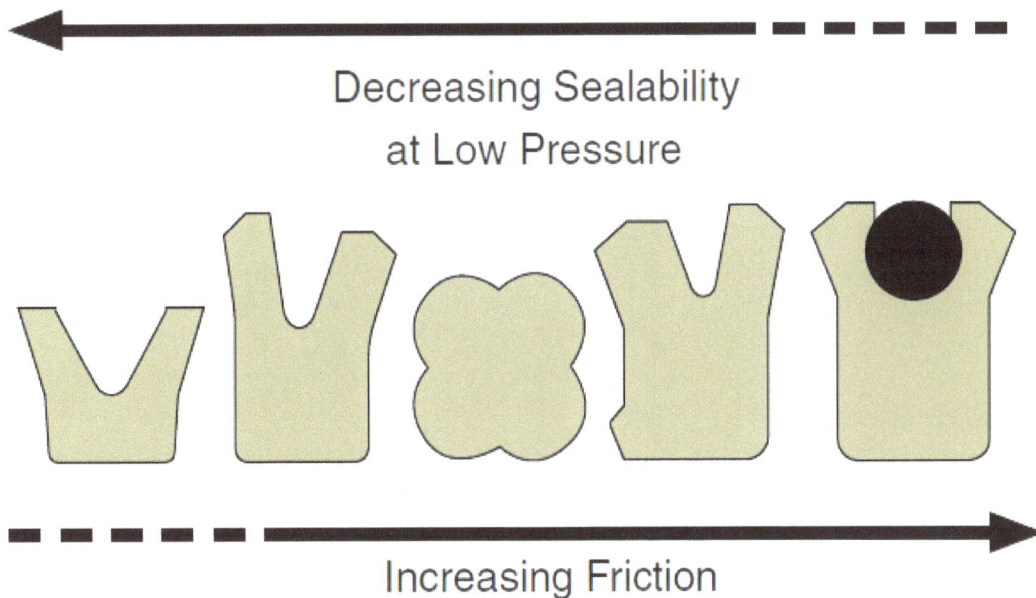

Fig. A.67 - Lip vs. Squeeze Sealing (Courtesy of Parker)

6.5- Selection of Seal Material Based on Working Pressure

Range of working pressure for a hydraulic seal must be checked during selection. The seal that is designed to work at 150 bar (2,175 psi) can last for 2 years. If it works at 200 bar (2,900 psi) it will last for 2 months. If it works at 350 bar (5076) it will last for 2 days.

6.6- Selection of Seal Material Based on Working Temperature

Performance of a hydraulic seal is highly affected by working temperature. Synthetic rubber can be formulated for continuous use at high or low temperatures, or for occasional short exposure to wide variations in temperature.

Figure A.68 shows the recommended working temperature range for common elastomeric materials. As shown in the figure, the most commonly used seal material (NBR) can work at temperature range from -50 to 100 °C (-58 to 212 °F). Silicon rubber (Viton Seals) can work at the most extended temperature ranging from -50 to 200 °C (-58 to 392 °F).

Temperature Effects on Sealing Elements

High Temperatures:
At a working temperature above 38°C (100°F), a combination of the following effects could happen:
- Some of the oil at the interface surface evaporates resulting in dry-running condition.
- Lack of lubrication will cause greatly accelerated seal wear.
- High temperature softens the seal body resulting in seal extrusion.
- Prolonged exposure to excessive heat causes permanent surface hardening.
- Thermal expansion of synthetic rubber makes it difficult to design the grooves for the seals.

Low Temperatures:
At low temperature environment, a combination of the following effects could happen:
- When cooled, elastomer compounds lose their elasticity and may be damaged due to insufficient elastic memory to overcome the seal shrinkage.
- At very low temperatures, seals are hardened, have glasslike brittleness, and may shatter if mechanically struck or hit with pressure spike.

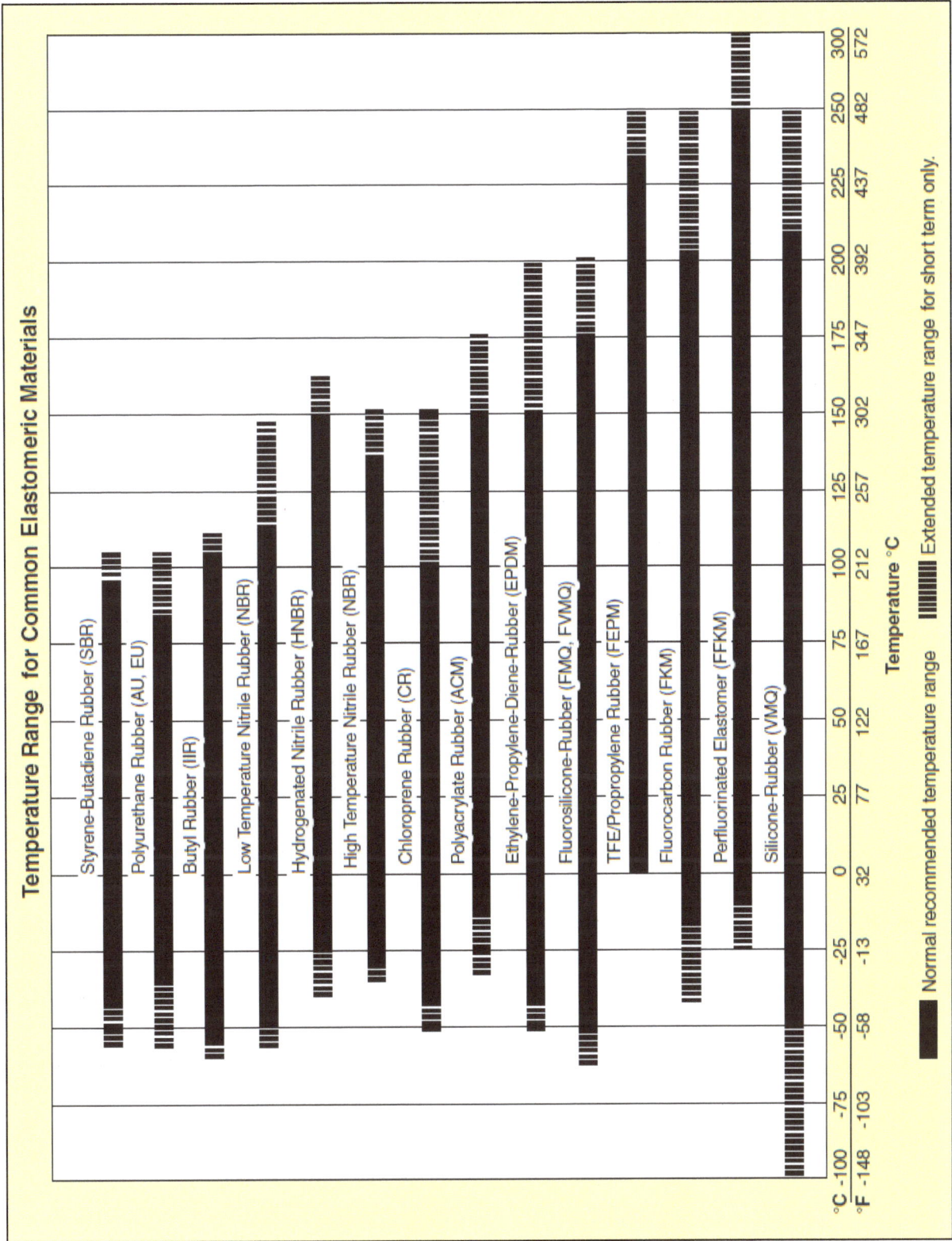

Fig A.68 - Recommended Working Temperature Range for Common Elastomeric Materials (Courtesy of Parker)

6.7- Selection of Seal Material Based on Working Fluid

Chemical compatibility is an important consideration when choosing the hydraulic fluid for the system. Hydraulic sealing elements must not react chemically with the working fluid.

Seal manufacturers should provide recommendations, as shown in Table A.6, for the best seal material for fire-resistant fluids.

Water also can adversely affect the seals, so special seals must be applied where water-based fire-resistant fluids are used. Table A.7 shows fluid compatibility for common sealing materials.

Properties of the Four Groups of Non-Flammable Fluids			
Properties	HFA/HFB	HFC	HFD
kinematic viscosity (mm²/s) to 50°C (122°F)	0.3 to 2	20 to 70	12 to 50
viscosity/temperature relationship	good	very good	bad
density at 15°C (59°F)	ca. 0.99	1.04 to 1.09	1.15 to 1.45
temperature range	3°C to 55°C (37°F to 131°F)	-25°C to 60°C (-13°F to 140°F)	-20°C to 150°C (-4°F to 302°F)
water content (weight %)	80 to 98	35 to 55	none
stability	emulsion poor solution very good	very good	very good
life of bearings	5 to 10%	6 to 15%	50 to 100%
heat transfer	excellent	good	poor
lubrication	acceptable	good	excellent
corrosion resistance	poor to acceptable	good	excellent
combustion temperature	not possible	after vaporizing of water under 1000°C (1832°F)	ca. 600°C (1112°F)
environmental risk	emulsion: used oil synth.: dilution	special waste	special waste
regular inspection	pH-level concentration water hardness micro-organisms	viscosity water content pH-level	viscosity neutral pH spec. gravity
seal material	NBR, FKM	NBR	FKM, EPDM[1]
(1) only for pure (mineral oil free) phosphate-ester (HFD-R)			

Table A.6 - Recommended Seal Materials for Fire-Resistant Fluids (Courtesy of Parker)

Seal Material	Compatible Fluids	Temperature Range
1. Metallic piston rings	Petroleum base and synthetic fluids, phosphate esters - for high pressure and severe conditions	Low to 500°F (260°C)
2. Leather	Petroleum base and some synthetics, phosphate esters - for medium to high pressure	-65°F to 225°F -54°C to 107°C
3. Neoprene rubber	General purpose industrial use, Freon™ 12; weather and salt water resistant	-65°F to 300°F -54°C to 149°C
4. Nitrile rubber (Buna N™)	Petroleum base fluids and mineral oils - used for some rotating seals, extrusion resistant	-65°F to 225°F -54°C to 107°C
5. Silicone rubber	Water and petroleum base fluids, phosphate esters; low tensile strength and tear resistance recommended for static seals only	-80°F to 450°F -62°C to 232°C
6. Fluoro-Elastomers (Viton™ and Fluorel™)	Petroleum base, synthetic, diester, silicate ester, and halogenated hydrocarbon fluids - for high temperature fluid applications	-20°F to 400°F -29°C to 204°C
7. Polyurethane	Petroleum base fluids - high resistance to ozone, sunlight and weathering; low water resistance	-65°F to 200°F -40°C to 93°C

**Table A.7 - Fluid Compatibility for Common Sealing Materials
(Hydraulic Specialist Study Manual, IFPS)**

6.8- Selection of Seal Material Based on Hardness

As it has been stated previously, while seal friction is inversely proportional to the seal hardness, leakage is directly proportional to seal hardness. Therefore, seal hardness should be compromised based on the seal application.

Example 1: Seal hardness should be high for Guide-Rings because their main objective is not to seal, but to guide a piston or a rod at low friction.

Example 2: Seal harness for piston seals should be lower for better sealability.

Figure A.69 shows that elastomers have the same trend that the seal hardness is inversely proportional, at different rates, with the working temperature. Various Nitrile compounds were used in this analysis.

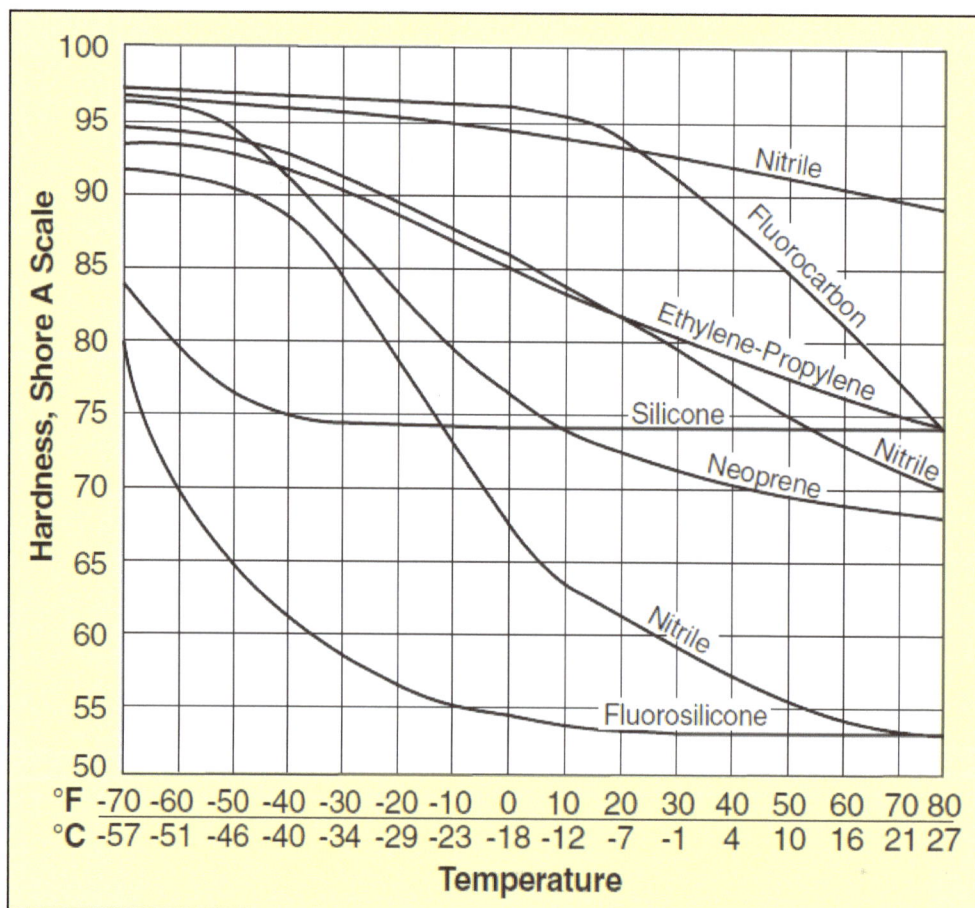

Fig A.69 - Effect of Low Temperature on Elastomer Hardness (Courtesy of Parker)

6.9- Selection of Seal Material Based on General Properties

The general properties of a hydraulic seal are identified based on the application to achieve a reasonable service life and optimum performance. Table A.8 shows the general properties of commonly known elastomers.

Comparison of Properties of Commonly Used Elastomers
(P = Poor – F = Fair – G = Good – E = Excellent)

Elastomer Type (Polymer)	Parker Compound Prefix Letter	Abrasion Resistance	Acid Resistance	Chemical Resistance	Cold Resistance	Dynamic Properties	Electrical Properties	Flame Resistance	Heat Resistance	Impermeability	Oil Resistance	Ozone Resistance	Set Resistance	Tear Resistance	Tensile Strength	Water/Steam Resistance	Weather Resistance
AFLAS (TFE/Prop)	V	GE	E	E	P	G	E	E	E	G	E	E	PF	PF	FG	GE	E
Butadiene		E	FG	FG	G	F	G	P	F	F	P	P	G	GE	E	FG	F
Butyl	B	FG	G	E	G	F	G	P	G	E	P	GE	FG	G	G	G	GE
Chlorinated Polyethylene		G	F	FG	PF	G	G	GE	G	G	FG	E	F	FG	G	F	E
Chlorosulfonated Polyethylene		G	G	E	FG	F	F	G	G	G	F	E	F	G	F	F	E
Epichlorohydrin	Y	G	FG	G	GE	G	F	FG	FG	GE	E	E	PF	G	G	F	E
Ethylene Acrylic	A	F	F	FG	G	F	F	P	E	E	F	E	G	F	G	PF	E
Ethylene Propylene	E	GE	G	E	GE	GE	G	P	G	G	P	E	GE	GE	GE	E	E
Fluorocarbon	V	G	E	E	PF	GE	F	E	E	G	E	E	E	F	GE	F	E
Fluorosilicone	L	P	FG	E	GE	P	E	G	E	P	G	E	G	P	F	F	E
Isoprene		E	FG	FG	G	F	G	P	F	F	P	P	G	GE	E	FG	F
Natural Rubber		E	FG	FG	G	E	G	P	F	F	P	P	G	GE	E	FG	F
Neoprene	C	G	FG	FG	FG	F	F	G	G	G	FG	GE	F	FG	G	F	E
HNBR	N, K	G	E	FG	G	GE	F	P	E	G	E	G	GE	FG	E	E	G
Nitrile or Buna N	N	G	F	FG	G	GE	F	P	G	G	E	P	GE	FG	GE	FG	F
Perfluorinated Fluoroelastomer	V, F	P	E	E	PF	F	E	E	E	G	E	E	G	PF	FG	GE	E
Polyacrylate	A	G	P	P	P	F	F	P	E	E	E	E	F	FG	F	P	E
Polysulfide		P	P	G	G	F	F	P	P	E	E	E	P	P	F	F	E
Polyurethane	P	E	P	FG	G	E	FG	P	F	G	G	E	F	GE	E	P	E
SBR or Buna S		G	F	FG	G	G	G	P	FG	F	P	P	G	FG	GE	FG	F
Silicone	S	P	FG	GE	E	P	E	F	E	P	FG	E	GE	P	P	F	E

Table A.8 - General Properties of Commonly Known Elastomers (Courtesy of Parker)

Chapter 7 – Sealing Solutions for Hydraulic Cylinders

7.1- Considerations for Hydraulic Cylinders Reliable Sealing

Hydraulic cylinders are required to operate leak-free in a variety of applications and environmental conditions, including exposure to abrasives, debris and both high and low temperatures.

Therefore, hydraulic seal manufacturers aggressively compete against each other to develop a comprehensive range of fluid power seals made from high quality materials with optimized designs to meet application requirements.

For a cylinder to perform reliably under leak-free conditions, the following set of features should be considered during designing a sealing package for a cylinder:
- Seals must be able to function at specified pressure and temperature.
- Seals must be able to withstand expected pressure spikes.
- Seals must be able to carry expected lateral loads.
- Seals must be compatible with type of hydraulic fluid used.
- Seals must offer conditions of low friction.
- Seals must be designed for easy installation and port passing.

To help cylinder seals to perform reliably:
- Cylinder must be mounted coaxially with the load to minimize the lateral force.
- Cylinder mounting points should be attached on a non-vibrating frame of references.
- Cylinder should operate in a clean environment.
- Cylinder should operate within the recommended working temperature and pressure.

There is no one sealing solution that is good for all cylinders. Design of a cylinder sealing package is a case-by-case and highly depends on the application and working conditions.

Table A.9 shows some working conditions and typical applications for hydraulic cylinders that should be taken into considerations when designing a sealing solution.

CYLINDER SPECIFICATION		LIGHT-DUTY		MEDIUM-DUTY		HEAVY-DUTY	
PRESSURE	Max	350 bar	5000 psi	500 bar	7500 psi	700 bar	10000 psi
	Normal Working	160 bar	2300 psi	250 bar	3625 psi	400 bar	5800 psi
		No pressure peaks		Intermittent pressure peaks		Regular pressure peaks	
Design		Lower operating stresses. Rigid well- aligned mounting, minimal side loading.		Steady operating stresses with intermittent high stress, some side loading.		Highly stressed for the majority of its working life. Side loading common.	
Condition of Fluid		Good system filtration. No cylinder contamination likely.		Good system filtration, but some cylinder contamination likely.		Contamination unavoidable from internal and external sources.	
Working Environment		Clean and inside a building. Operating temperature variations limited.		Mixture of indoors and outdoors but some protection from the weather.		Outdoors all the time or dirty indoor area. Wide variations in temperature, both ambient and working. Difficult service conditions.	
Usage		Irregular with short section of stroke at working pressures. Regular usage but at low pressure.		Regular usage with most of the stroke at working pressure.		Large amount of usage at high pressure with peaks throughout the stroke.	
Typical Applications		Machine tools Lifting equipment Mechanical handling Injection moulding machines Control and robot equipment Agricultural machinery Packaging equipment Aircraft equipment Light duty tippers		Heavy duty lifting equipment Agricultural equipment Light duty off-road vehicles Cranes and lifting platforms Heavy duty machine tools Injection moulding machines Some auxiliary mining machinery Aircraft equipment Presses Heavy duty tippers (telescopic) Heavy duty mechanical handling		Foundry and metal fabrication plant Mining machinery Roof supports Heavy duty earthmoving machinery Heavy duty off-road vehicles Heavy duty presses	

**Table A.9 – Working Conditions and Typical applications for Hydraulic Cylinders
(Courtesy of Hallite Seals)**

Figure A.70 shows the main components of a typical sealing solution for a light-duty hydraulic cylinder. As shown in the figure, piston seal package contains a piston dynamic seal and a Wear or a Guide Ring. Piston static seal is used if the piston head is assembled with the piston rod.

Like pistons, cylinder rod seal package contains a dynamic seal and a Guide Ring. Unlike pistons, cylinder rod has additional sealing elements such as Rod Wiper and a Buffer Seal. A static seal is used at the rod side to seal the cylinder head against the barrel to prevent external leakage.

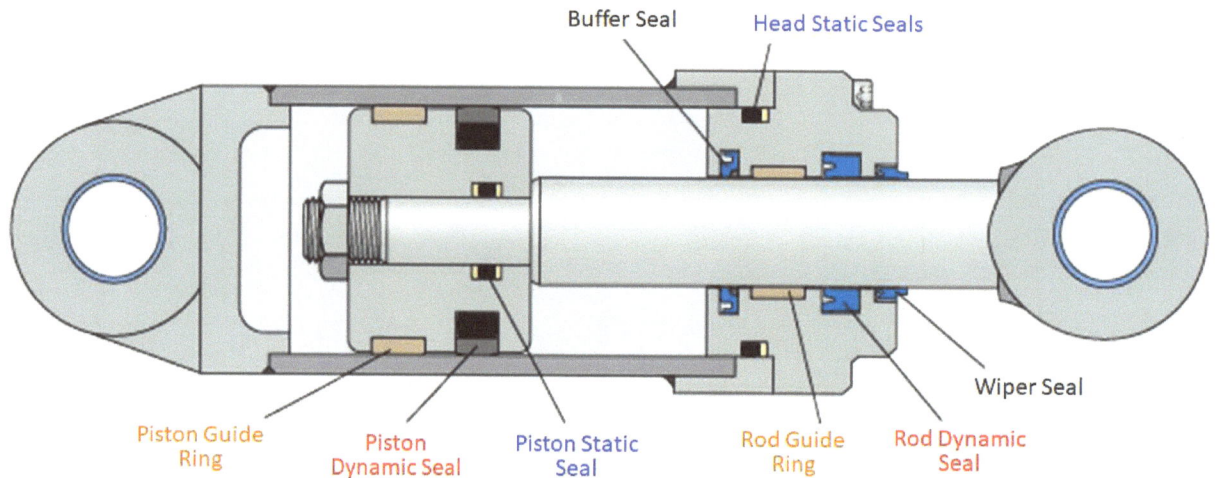

Fig. A.70 - Basic Components of Cylinder Sealing Solutions

The following sections overview the common piston and rod sealing solutions. Presented examples cover cylinders for light-duty applications [pressure up to 350 bar (5000 psi)], medium-duty applications [pressure up to 500 bar (7,500 psi)], high-duty applications (pressure up to 700 bar (10,000 psi)]. Piston and rod seals are generally classified as *Unidirectional* (*Single-Acting*) and *Bidirectional* (*Double-Acting*) seals.

7.2- Sealing Solutions for Cylinder Rods

7.2.1- Unidirectional Rod Sealing Solutions

Rod seals must exhibit no dynamic leakage to the atmosphere side under all operating conditions and must be completely leak-free when the cylinder rod is at a standstill. The following rod sealing solutions are good to seal cylinder rods against external leakage to the atmospheric side.

Example 1: U-Cup Seals for Unidirectional Rod Sealing

Figure A.71 shows various types of U-Cup seals for cylinder rod sealing as follows

Type 1: The U-Cup has two sealing lips in the dynamic sealing zone. The two sealing lips provide improved sealing at low system pressures. Furthermore, the second sealing lip prevents entry of dirt from the atmospheric side.

Type 2: The O-Ring loaded U-Cup seal has stronger radial contact forces with increased system pressure. Hence, it has excellent sealing behavior with and without pressure activation. The short sealing lip reduces friction compared to common U-Cups.

Type 3: The spring energized U-Cup seal has the same advantage as type 2 plus it is more dynamically stable.

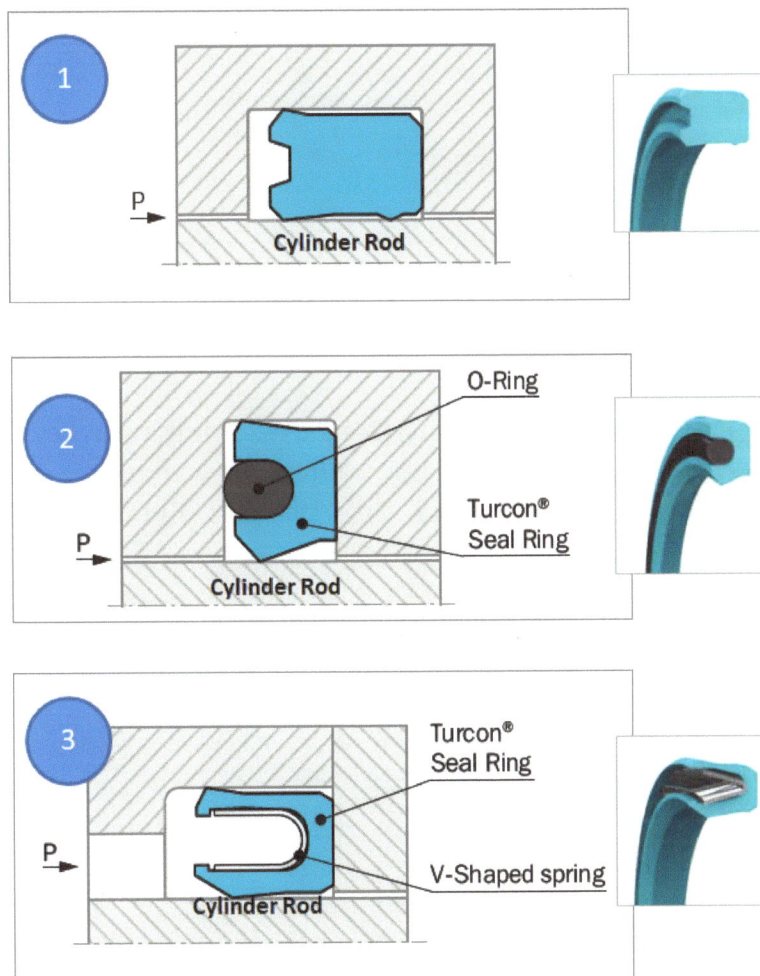

Fig. A.71 - U-Cup Seals for Unidirectional Rod Sealing (Courtesy of Trelleborg)

Example 2: V-Packing Seals for Unidirectional Rod Sealing
This rod sealing package, as shown in Fig. A.72, utilizes a set of V-Packings as the dynamic seal. This solution is recommended for applications with extremely high-pressure and possible pressure spikes.

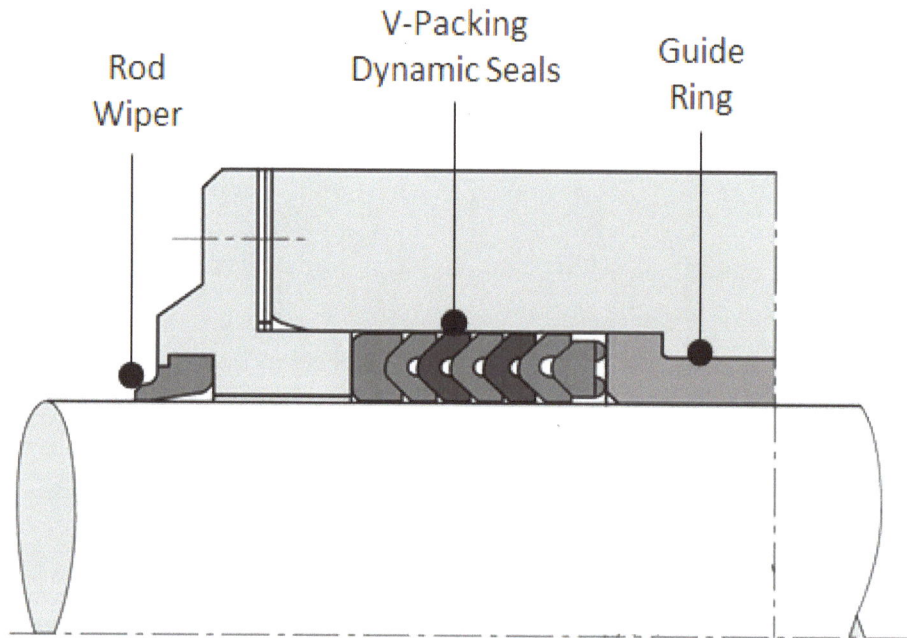

Fig. A.72 - V-Packing Seals for Unidirectional Rod Sealing

Example 3: Low Friction Rod Seals for Better Positioning Accuracy
Figure A.73 shows a single-acting rod seal where high demands are made on positional accuracy and stick-slip-free movement, e.g. closed-loop position-controlled servo cylinders).

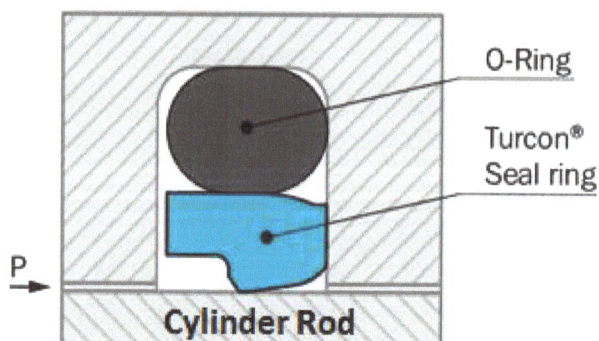

**Fig. A.73 - Unidirectional Rod Sealing Solution for Better Positioning Accuracy
(Courtesy of Trelleborg)**

Example 4: Cylinder Rod Unidirectional Sealing Solution for Light-Duty Applications
This rod sealing package, as shown in Fig. A.74, contains a single U-Cup Seal to prevent external leakage. The U-Cup seal can be *O-Ring-loaded* or *Spring-Energized.* The sealing package also contains a Rod-Wiper to remove dirt from the rod surface during retraction.

The figure shows a Guide-Ring to carry lateral forces. As shown in the figure, the Guide-Ring position changes depending on the lateral force. The Guide-Ring should be closer to the wiper as the lateral force increases. As shown in Fig. A.75, to improve the operational safety under high lateral loads, up to three Wear-Rings can be installed.

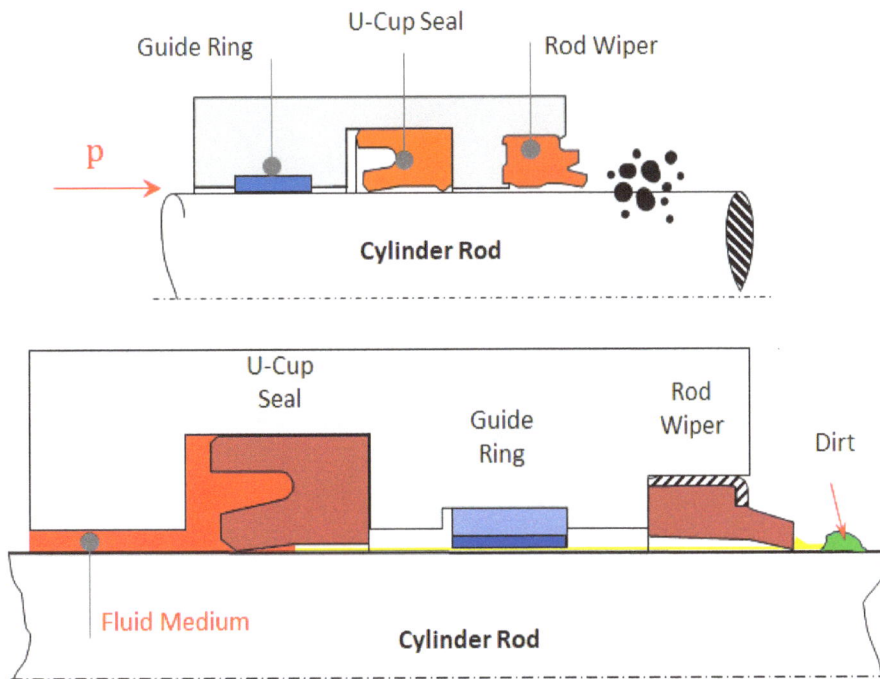

Fig. A.74- Cylinder Rod Unidirectional Sealing Solution for Light-Duty Applications (Courtesy of Trelleborg)

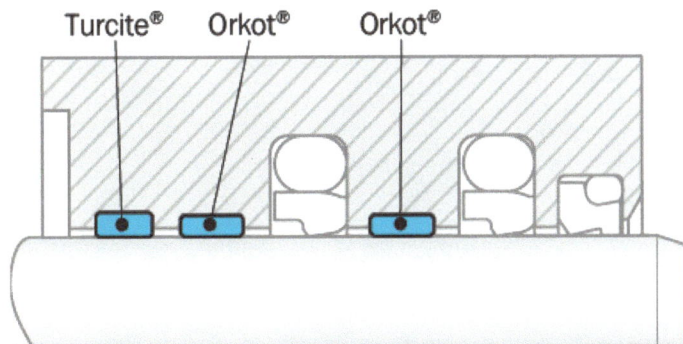

Fig. A.75- Wear-Rings for Cylinder Designs

Example 5: Cylinder Rod Unidirectional Sealing Solution for Medium-Duty Applications
This rod sealing package, as shown in Fig. A.76, contains two dynamic seals, a primary seal and a low-pressure U-Cup secondary seal. When two dynamic seals are used, they are referred to as a *Redundant Sealing System*. Obviously, the package contains a Rod Wiper and a Guide-Ring. Figure A.77 shows an example of that same rod sealing solution but from a different manufacturer.

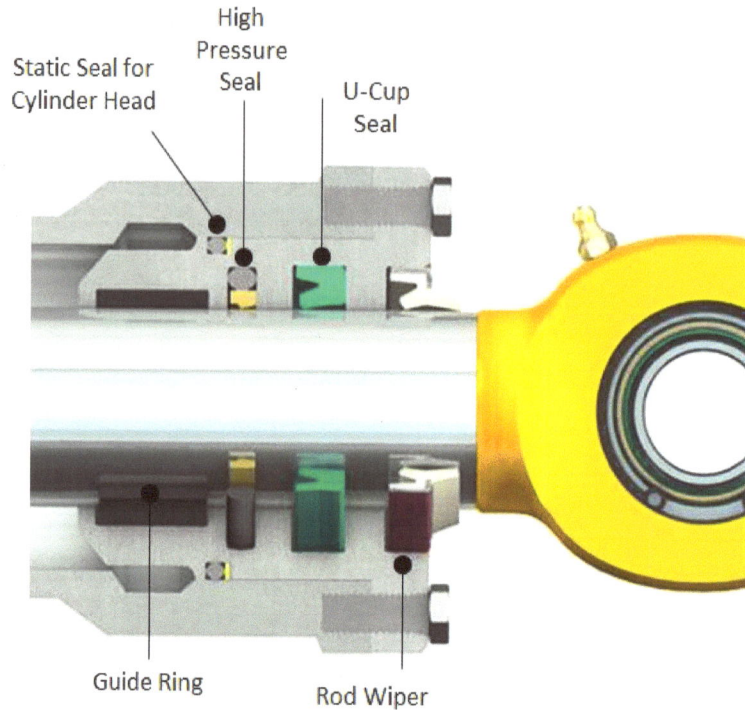

Fig. A.76 - Cylinder Rod Unidirectional Sealing Solution for Medium-Duty Applications
(www.skf.com)

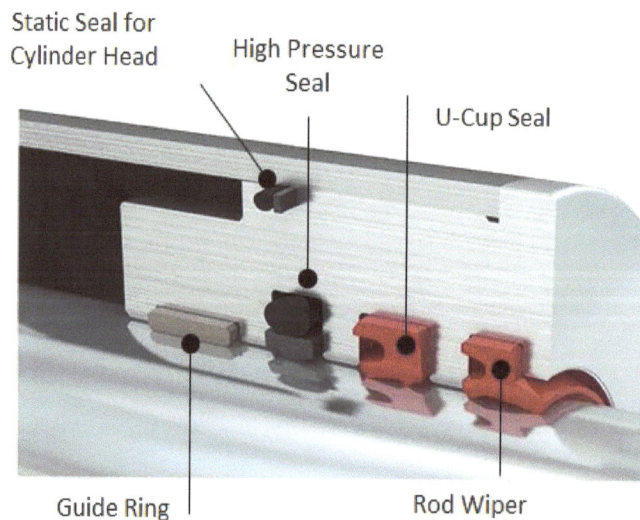

Fig. A.77 - Cylinder Rod Unidirectional Sealing Solution for Medium-Duty Applications
(Courtesy of American High-Performance Seals)

Example 6: Redundant Unidirectional Rod Sealing Solution for Heavy-Duty Applications
This rod sealing package, as shown in Fig. A.78, contains a *Redundant* sealing system. It is also referred to as "*Double Sealing*" or "*Tandem Sealing*" solution. As shown in the figure, the seal is constructed from an O-Ring for static sealing and a custom-designed seal for dynamic sealing.

Redundant sealing systems are used where:
- Environmentally harmful fluids are used.
- A single seal can't withstand the application conditions over the demanded service life.
- A machine starts at low temperature (*Cold Starts*). During cylinder extension, due to the very high viscosity, the oil is heated because the friction at the primary seal and is then reliably wiped off - at a now lower viscosity - by the secondary seal.

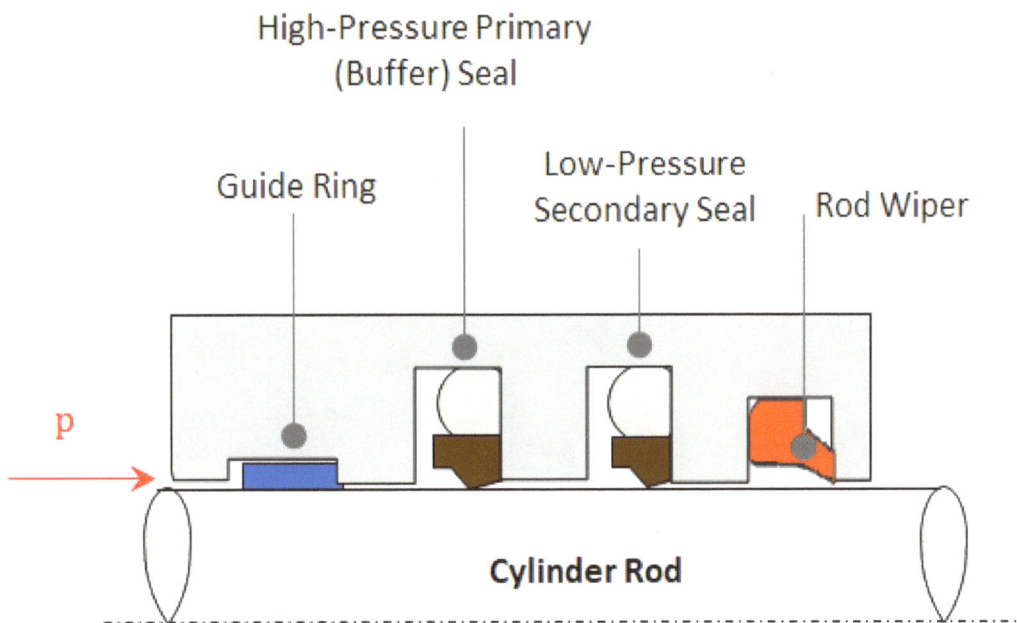

Fig. A.78 - Redundant Unidirectional Rod Sealing Solution for Heavy-Duty Applications

Figure A.79 shows a typical industry example for a redundant sealing solution. During cylinder retraction, the oil is stored in the reservoir between the primary and the secondary seals. As the rod continues to retract, the oil is pumped back hydrodynamically through the primary seal clearance against the system pressure. This design is known as the *Back-Pumping Effect*.

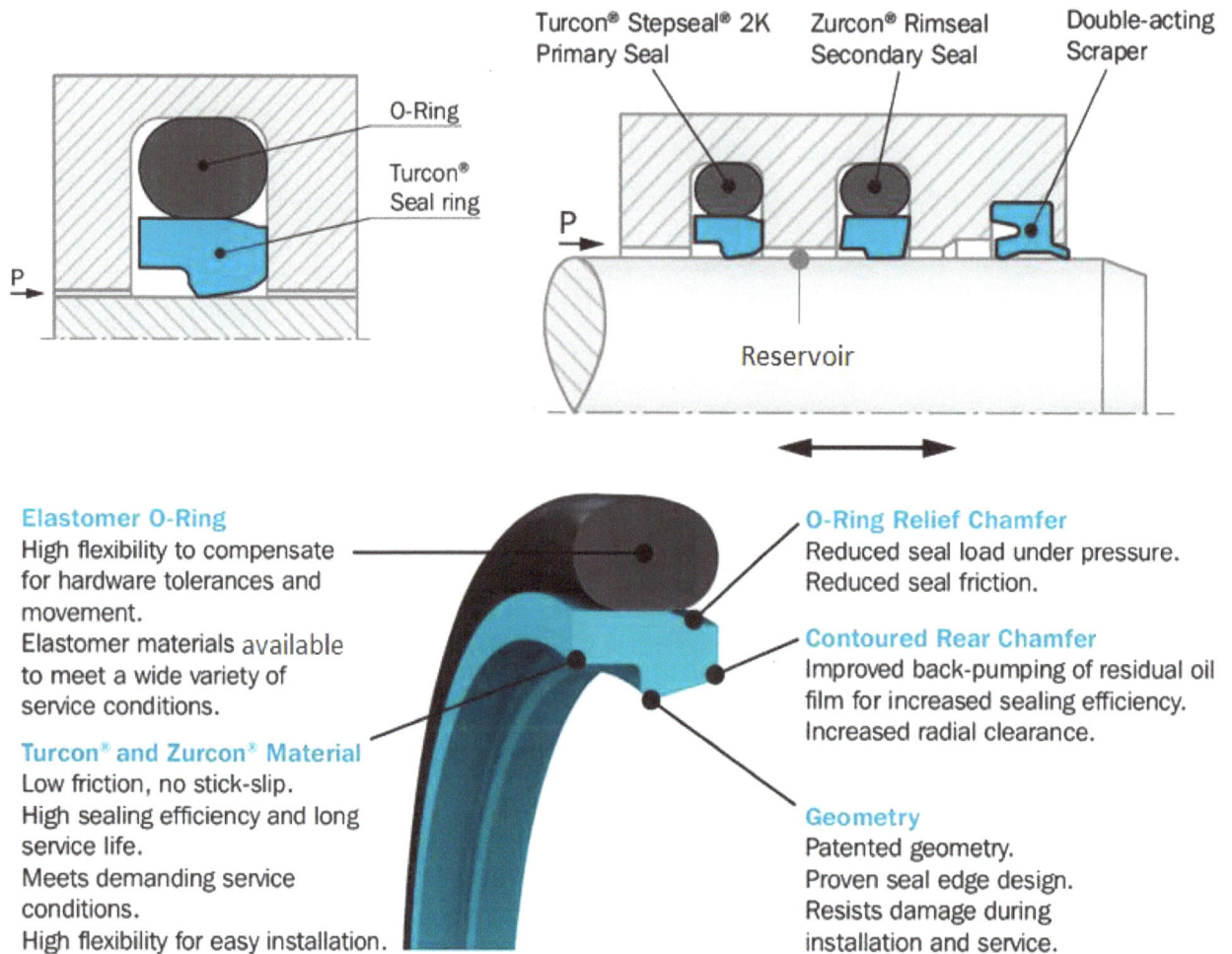

Elastomer O-Ring
High flexibility to compensate for hardware tolerances and movement.
Elastomer materials available to meet a wide variety of service conditions.

Turcon® and Zurcon® Material
Low friction, no stick-slip.
High sealing efficiency and long service life.
Meets demanding service conditions.
High flexibility for easy installation.

O-Ring Relief Chamfer
Reduced seal load under pressure.
Reduced seal friction.

Contoured Rear Chamfer
Improved back-pumping of residual oil film for increased sealing efficiency.
Increased radial clearance.

Geometry
Patented geometry.
Proven seal edge design.
Resists damage during installation and service.

Fig. A.79 - Typical Redundant Unidirectional Rod Sealing Solution for Heavy-Duty Applications (Courtesy of Trelleborg)

Example 7: Redundant Unidirectional Sealing Solution for Slow and Long Stroke Cylinder Rods

In long-stroke cylinders and equipment operating with low speed during retraction, it has been found that hydrodynamic back-pumping may become insufficient to prevent build-up of pressure in the seal system behind the primary seal. Pressure build-up in the seal system leads to leakage, increased friction and wear, and may ultimately require replacement of the seals. The usual solution in such equipment has been to provide space for a buffer volume behind the primary seal or to install a drain line. As shown in Fig. A.80, first invented by *Trelleborg Sealing Solutions*, the built-in check valve function eliminates pressure build-up and so prevent pressure built up in the reservoir volume. Hence, improve sealing performance with outstanding sliding and wear resistance properties.

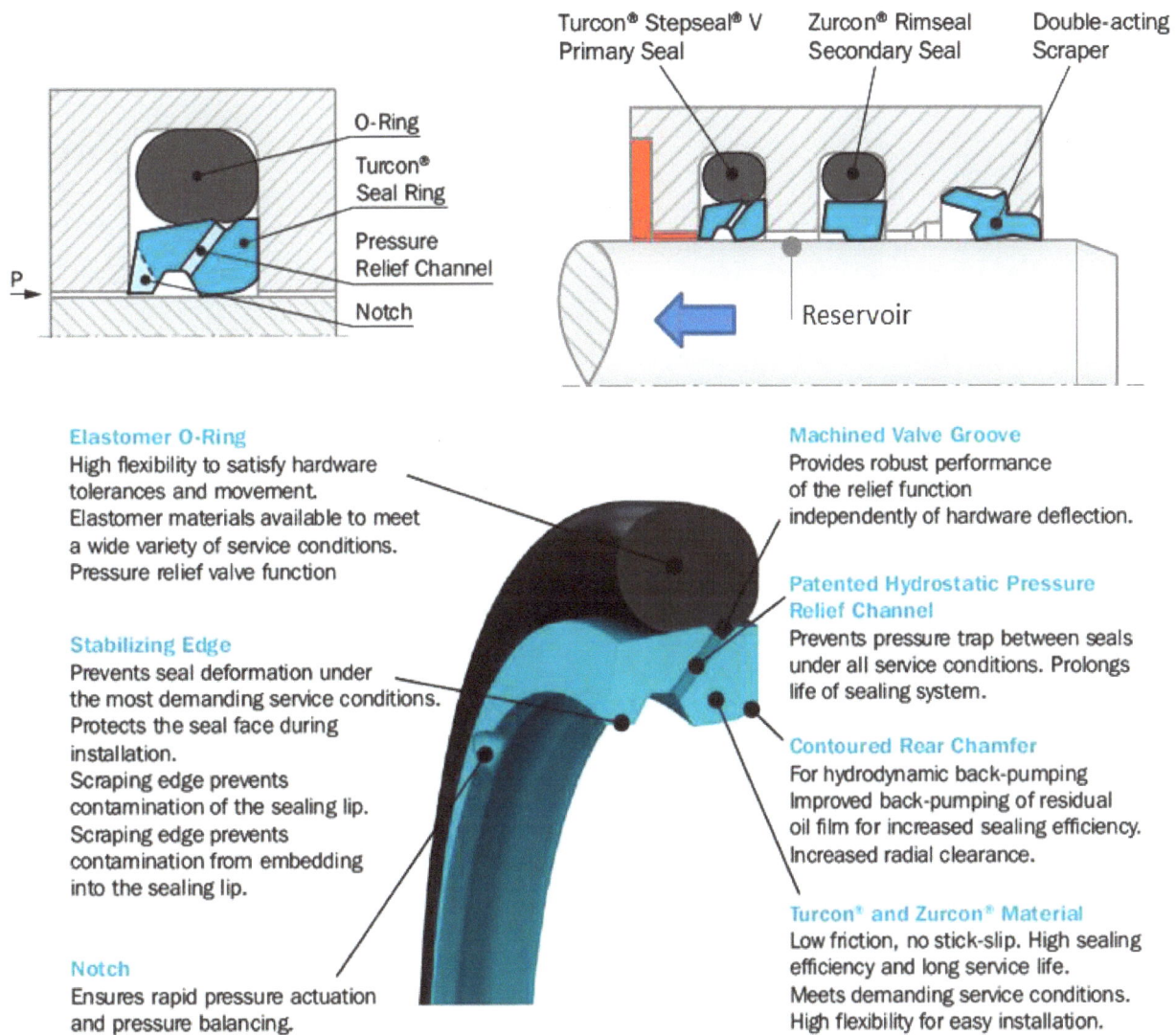

Turcon® Stepseal® V Primary Seal

Zurcon® Rimseal Secondary Seal

Double-acting Scraper

O-Ring

Turcon® Seal Ring

Pressure Relief Channel

Notch

P

Reservoir

Elastomer O-Ring
High flexibility to satisfy hardware tolerances and movement.
Elastomer materials available to meet a wide variety of service conditions.
Pressure relief valve function

Stabilizing Edge
Prevents seal deformation under the most demanding service conditions.
Protects the seal face during installation.
Scraping edge prevents contamination of the sealing lip.
Scraping edge prevents contamination from embedding into the sealing lip.

Notch
Ensures rapid pressure actuation and pressure balancing.

Machined Valve Groove
Provides robust performance of the relief function independently of hardware deflection.

Patented Hydrostatic Pressure Relief Channel
Prevents pressure trap between seals under all service conditions. Prolongs life of sealing system.

Contoured Rear Chamfer
For hydrodynamic back-pumping Improved back-pumping of residual oil film for increased sealing efficiency.
Increased radial clearance.

Turcon® and Zurcon® Material
Low friction, no stick-slip. High sealing efficiency and long service life.
Meets demanding service conditions.
High flexibility for easy installation.

Fig. A.80 - Redundant Unidirectional Sealing Solution for Slow and Long Stroke Cylinder Rods (Courtesy of Trelleborg)

Example 8: Redundant Unidirectional Rod Sealing Solution for Heavy-Duty and Large Lateral Force Applications

Figure A.81 shows a typical industry example for a *Redundant* (*Tandem*) sealing system developed to meet heavy duty demands. For large lateral forces, two Guide-Rings are used.

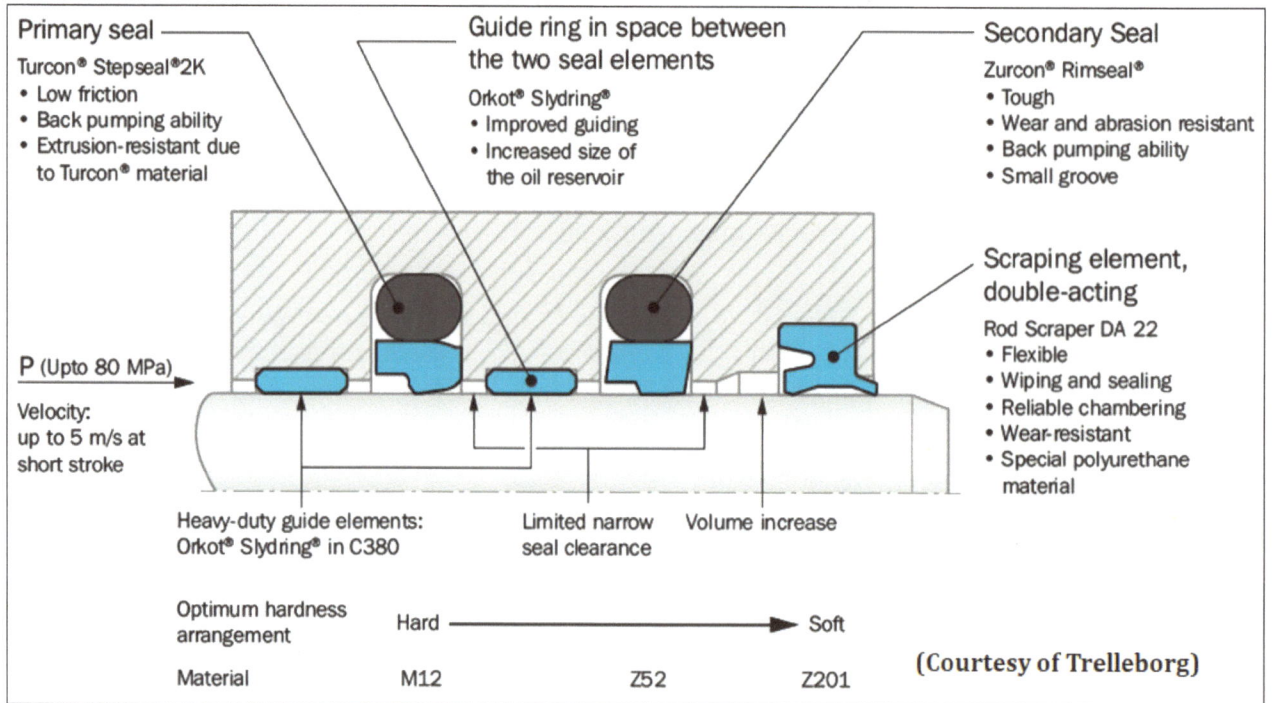

Primary seal

Turcon® Stepseal®2K
- Low friction
- Back pumping ability
- Extrusion-resistant due to Turcon® material

Guide ring in space between the two seal elements

Orkot® Slydring®
- Improved guiding
- Increased size of the oil reservoir

Secondary Seal

Zurcon® Rimseal®
- Tough
- Wear and abrasion resistant
- Back pumping ability
- Small groove

Scraping element, double-acting

Rod Scraper DA 22
- Flexible
- Wiping and sealing
- Reliable chambering
- Wear-resistant
- Special polyurethane material

P (Upto 80 MPa)

Velocity:
up to 5 m/s at short stroke

Heavy-duty guide elements:
Orkot® Slydring® in C380

Limited narrow seal clearance

Volume increase

Optimum hardness arrangement   Hard ——————————————➤ Soft

Material   M12   Z52   Z201

(Courtesy of Trelleborg)

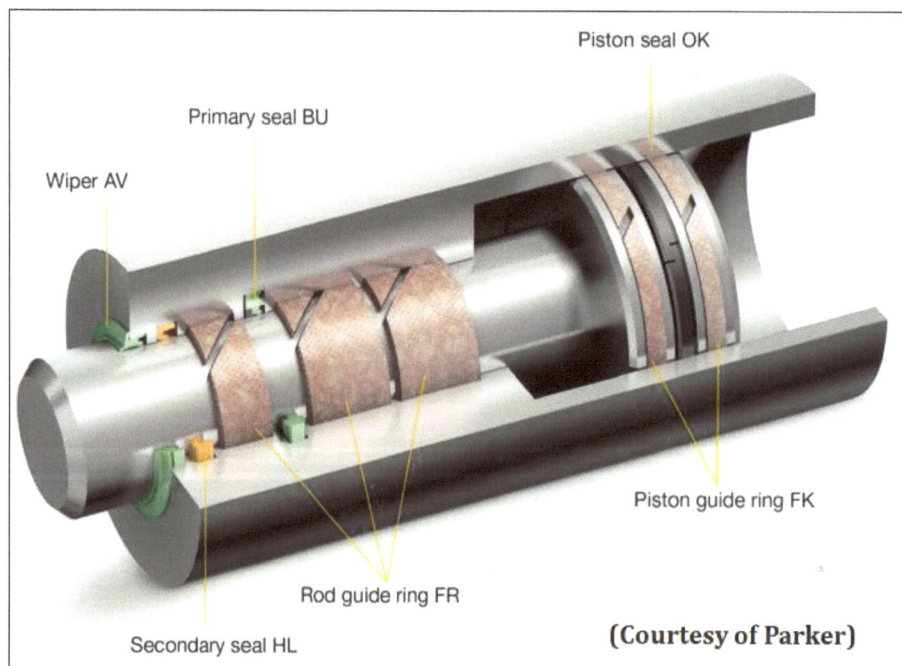

Piston seal OK

Primary seal BU

Wiper AV

Piston guide ring FK

Rod guide ring FR

Secondary seal HL

(Courtesy of Parker)

Fig. A.81 - Redundant Unidirectional Rod Sealing Solution for Heavy-Duty and Large Lateral Force Applications

7.2.2- Bidirectional Rod Sealing Solutions

Example 1: Cylinder Rod Basic Bidirectional Sealing Solution for Light-Duty Applications
In a tandem cylinder shown in Fig. A.82, the double-acting seal is a combination of a custom sealing ring and an energizing O-Ring. Together, with the squeeze of the O-Ring, ensures a good sealing effect even at low pressure. At higher system pressures, the O-Ring is energized by the fluid, pushing the sealing ring face with increased force.

As shown in the figure, the trapezoidal profile cross section of the custom sealing ring allows the lubricating hydrodynamic fluid film to be built under the seal with stick-slip-free linear motion.

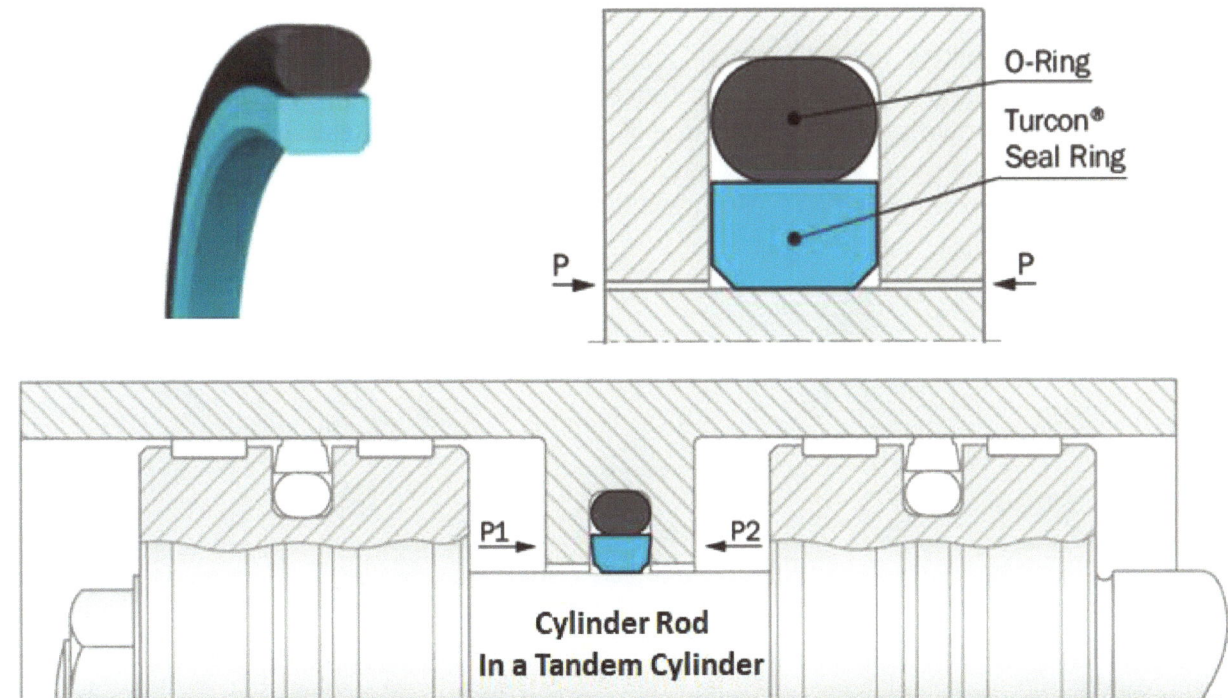

**Fig. A.82 - Cylinder Rod Basic Bidirectional Sealing Solution for Light-Duty Applications
(Courtesy of Trelleborg)**

Example 2: Bidirectional Rod Sealing Enhanced Solution for Medium-Duty Applications
Figure A.83 shows various enhancements for the basic bidirectional rod seal design.

Solution 1 for Blow-By-Effect: shows one radial notch is made on each side of the seal ring. The notches override the *Blow-By-Effect*. The notches are made to assure that a rapid energizing of the seal takes place at sudden changes in pressure and direction of motion.

Solution 2 for short-stroke and high-frequency: Reciprocating movement, with an increasing frequency above 5 Hz, the formation of lubrication under the contact face of the seal ring is reduced. High-frequency is most often occurring in connection with short-strokes. These two types of movements together accelerate the wear on hardware and seal. In this solution a symmetric seal is used with angled contact faces to ensure oil film is not scraped away from the surface. Oil is transported into the groove in the middle of the contact area forming an oil reservoir for lubrication. Wear particles are also likely to be captured in this groove, thus preventing them from embedding in the surface where the highest contact force occurs.

Solution 3 for better leakage control: The seal is designed with two seal edges. It acts as the primary seal for pressures from both sides, prevents build-up of hydrodynamic pressure over the seal profile, and prevents the risk of blow-by effect. The central sealing face increases the sealing effect.

Solution 4 for better leakage control and low friction: The sealing ring and the *Bean Seal* together create the dynamic sealing function while the O-Ring performs the static sealing function. The shown design incorporates a limited foot print Bean Seal in the dynamic sealing face. This optimizes leakage control while minimizing friction.

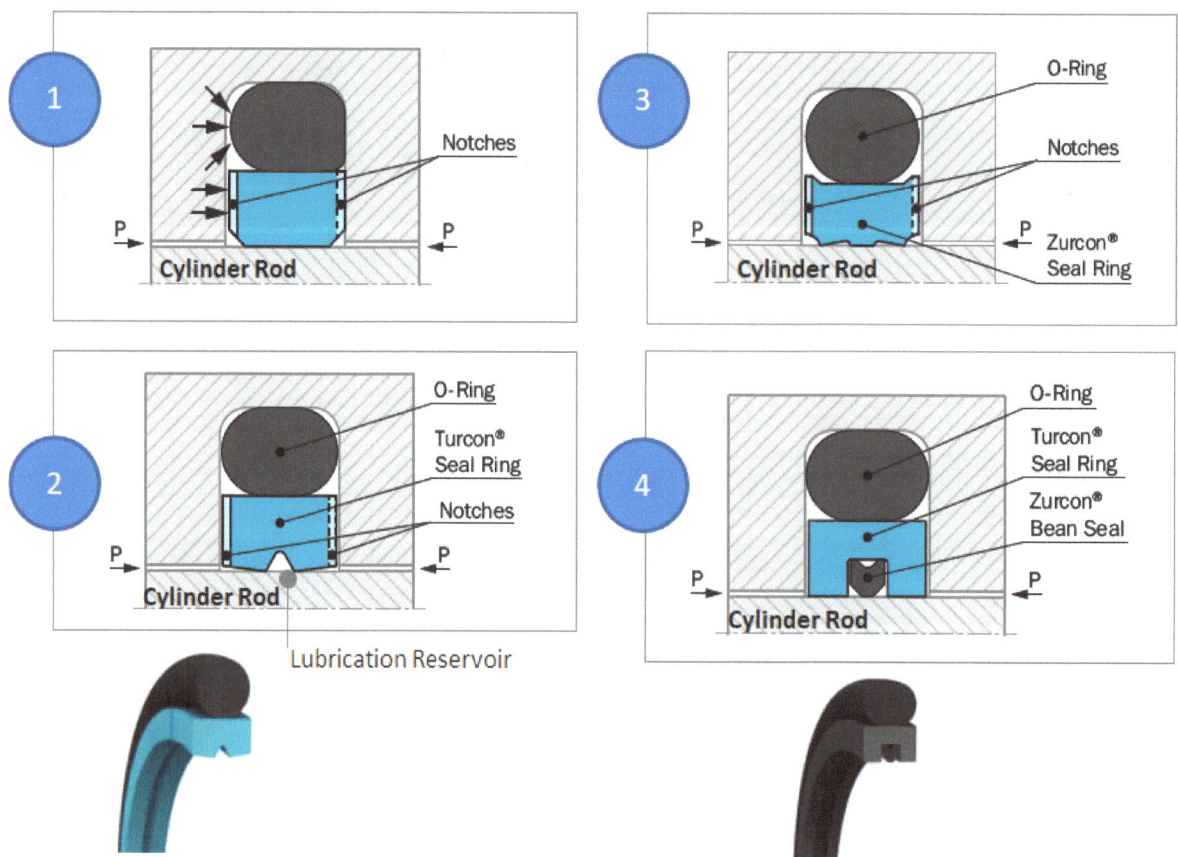

**Fig. A.83 - Enhanced Bidirectional Rod Seal Designs for Medium-Duty Applications
(Courtesy of Trelleborg)**

Example 3: U-Cup Seals for Bidirectional Rod Sealing
Figure A.84 shows two single-acting U-cup seals which are placed back-to-back to form bidirectional rod sealing solution. As shown in the figure, U-Cup seals could be O-Ring loaded or spring-loaded.

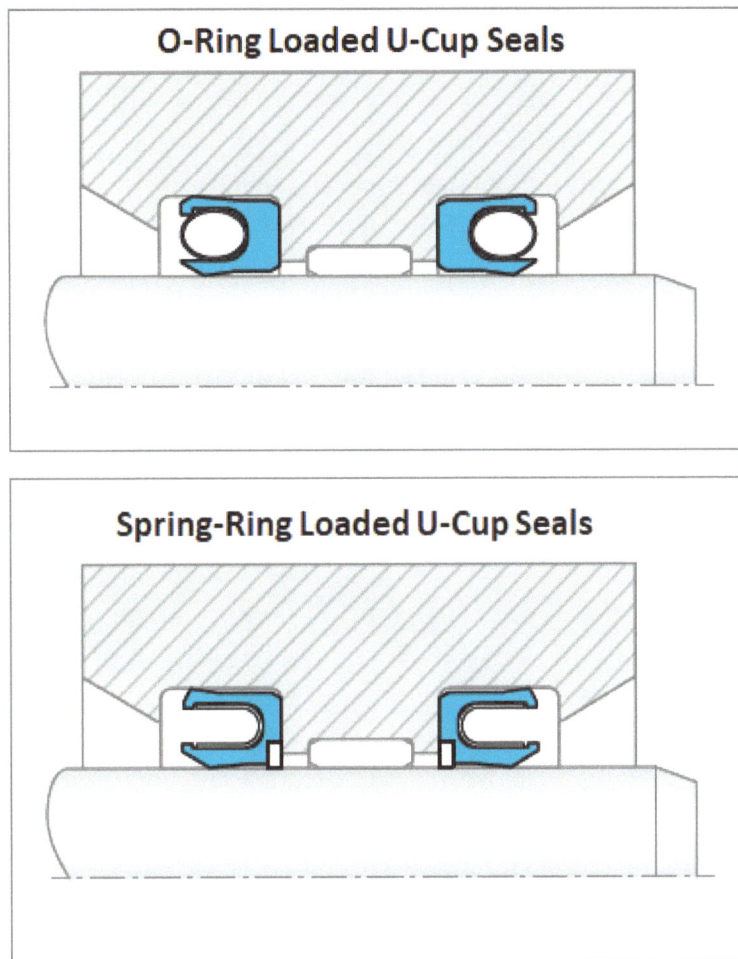

**Fig. A.84 - Cylinder Rod Bidirectional Sealing Solutions using U-Cup Seals
(Courtesy of Trelleborg)**

7.3- Sealing Solutions for Cylinder Pistons

7.3.1- Unidirectional Piston Sealing Solutions

The following piston sealing solutions are good for cylinders that develop power during extension stroke and retracts load-free at low pressure.

Example 1: U-Cup Seals for Unidirectional Piston Sealing

Figure A.85 shows various types of U-Cup seals for cylinder pistons seal as follows

Type 1: The U-Cup shown in the figure is provided with a robust dynamic sealing lip and a wide contact area of the static lip, which guaranties an effective position in the groove. The profile is suitable for pressures up to 400 bar (5,801 psi) provided that the extrusion gap is adapted to the pressure level.

Type 2: The O-Ring loaded U-Cup seal provides static sealing by the O-Ring. The O-Ring is protected from damage under high pressure and pressure cycles by the contoured O-Ring contact zone which supports the O-Ring. The seal shown in the figure is designed with hydrodynamic back-pumping effect which allows the seal to relieve pressure trapped between seals in tandem configurations.

Type 3: The spring energized U-Cup seal is a more dynamically stable.

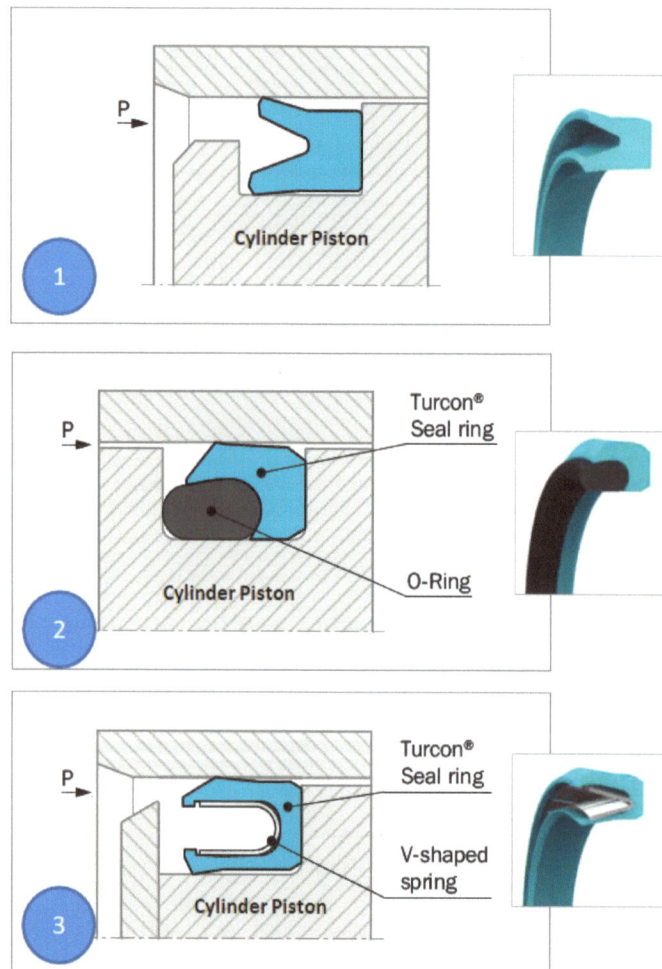

Fig. A.85 - U-Cup Seals for Unidirectional Piston Sealing (Courtesy of Trelleborg)

Example 2: V-Packing Seals for Unidirectional Piston Sealing

This piston seal, as shown in Fig. A.86, utilizes a set of V-Packings as the dynamic seal. This solution is assumed for applications with high extremely high-pressure and possible pressure spikes.

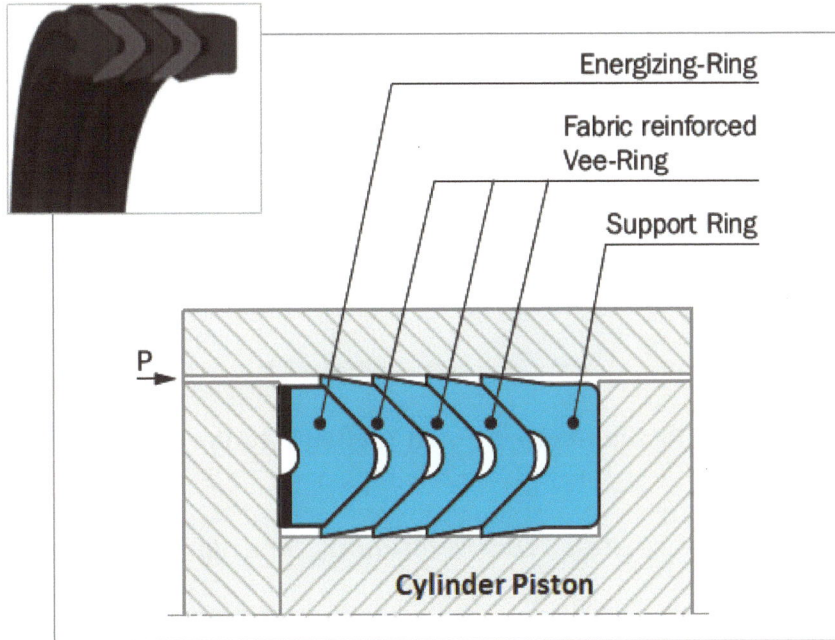

Fig. A.86 - V-Packing Seals for Unidirectional Piston Sealing (Courtesy of Trelleborg)

Example 3: Piston Seals for Better Positioning Accuracy

Figure A.87 shows a single-acting piston seal where high demands are made on positional accuracy and stick-slip-free movement.

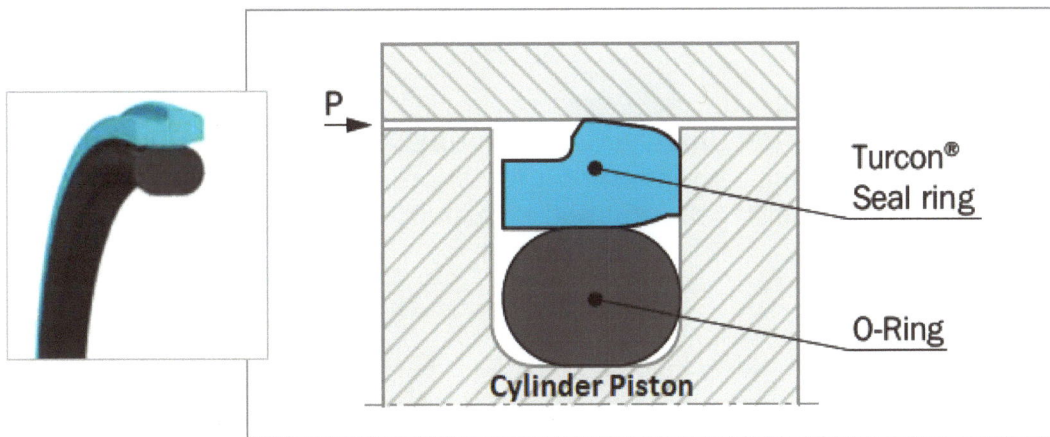

Fig. A.87 - Unidirectional Piston Sealing Solution for Better Positioning Accuracy (Courtesy of Trelleborg)

Example 4: Cylinder Piston Unidirectional Sealing Solution for Light-Duty Applications
This piston seal package, as shown in Fig. A.88, contains a single U-Cup Seal to prevent external leakage. The U-Cup seal can be *O-Ring-loaded* or *Spring-Energized.* A piston guide ring offers bearing surface for lateral load.

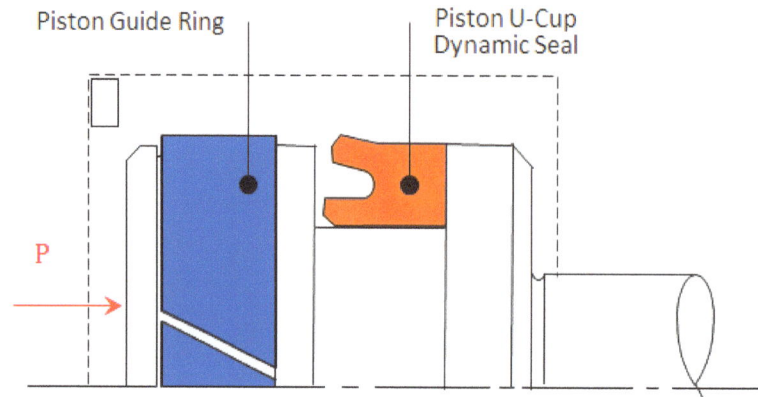

Fig. A.88 - Cylinder Piston Unidirectional Sealing Solution for Light-Duty Applications

Example 5: Cylinder Piston Unidirectional Sealing Solution for Medium-Duty Applications
As shown in Fig. A.89, for medium-high pressure applications, unidirectional piston U-Cup seals are O-Ring loaded or Spring-Energized
.

Fig. A.89 - Cylinder Piston Unidirectional Sealing Solution for Medium-Duty Applications
(Courtesy of American High-Performance Seals)

Example 6: Cylinder Piston Unidirectional Sealing Solution for Heavy-Duty Applications
As shown in Fig. A.90, for high pressure applications, unidirectional piston U-Cup seals are supported by Back-up Rings to prevent seal extrusion.

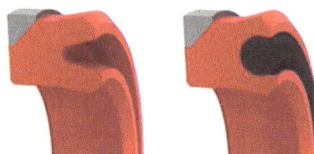

Fig. A.90 - Cylinder Piston Unidirectional Sealing Solution for Heavy-Duty Applications
(Courtesy of American High-Performance Seals)

Example 7: Redundant Unidirectional Sealing Solution for Slow and Long Stroke Cylinder Pistons

In long-stroke cylinders and equipment operating with low speed during retraction, it has been found that hydrodynamic back-pumping may become insufficient to prevent build-up of pressure in the seal system behind the primary seal. Pressure build-up in the seal system leads to leakage, increased friction and wear, and may ultimately require replacement of the seals. The usual solution in such equipment has been to provide space for a buffer volume behind the primary seal or to install a drain line. As shown in Fig. A.91, first invented by *Trelleborg Sealing Solutions*, the built-in check valve function eliminates pressure build-up in the reservoir volume. Hence, improving seal performance with outstanding sliding and wear resistance properties.

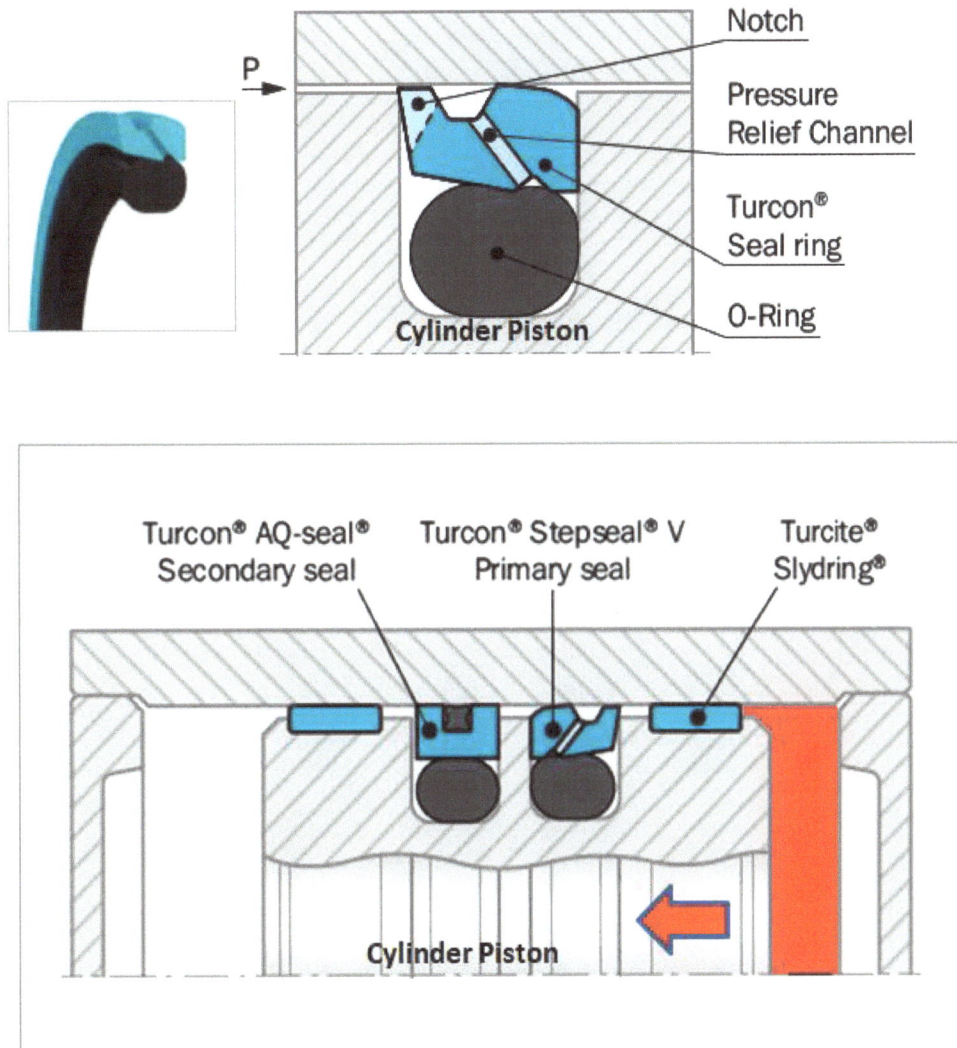

Fig. A.91 - Redundant Unidirectional Sealing Solution for Slow and Long Stroke Cylinder Rods (Courtesy of Trelleborg)

7.3.2- Bidirectional Piston Sealing Solutions

The following sealing solutions are good for cylinders that develops power during both extension and retraction directions.

Example 1: Basic Bidirectional Piston Sealing Solution for Light-Duty Applications
As shown in Fig. A.92, the double-acting seal is a combination of a custom seal ring and an energizing O-Ring. Together, with the squeeze of the O-Ring, ensures a good sealing effect even at low pressure.

At higher system pressures, the O-Ring is energized by the fluid, pushing the seal ring face with increased force. As shown in the figure, the trapezoidal profile cross section of the seal ring allows the lubricating hydrodynamic fluid film to be built above the seal with stick-slip-free linear motion. Two Guide-Rings are used to absorb lateral forces. Figure A.93 shows a typical industry example of a piston with such seal package.

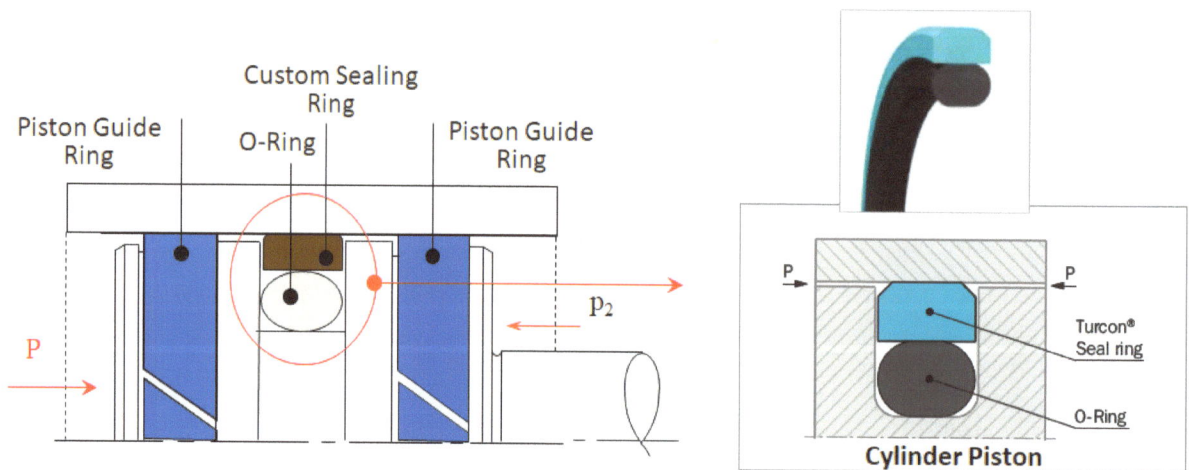

Fig. A.92 - Basic Bidirectional Piston Sealing Solution (Courtesy of Trelleborg)

Fig. A.93 - Basic Bidirectional Piston Sealing Solution for Light-Duty Applications

Example 2: Bidirectional Piston Sealing Enhanced Solution for Medium-Duty Applications

Figure A.94 shows a typical example from industry of a bidirectional piston sealing solution with an enhanced sealing ring.

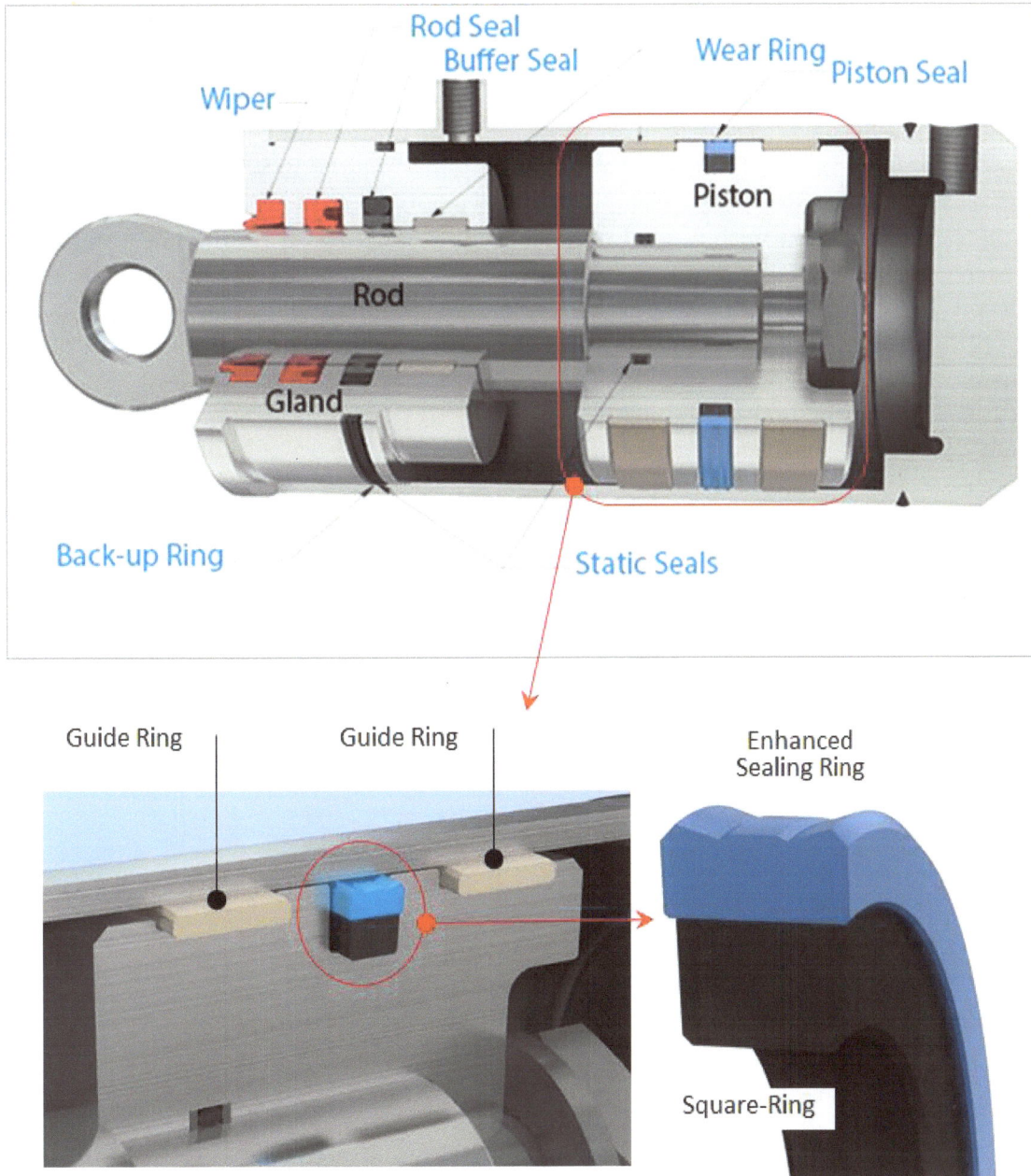

Fig. A.94 - Enhanced Bidirectional Piston Sealing Solution for Medium-Duty Applications (Courtesy of American High-Performance Seals)

Figure A.95 shows various enhancements for the basic bidirectional piton seal design.

Solution 1 for Blow-By-Effect: shows one radial notch is made on each side of the sealing ring. The notches override the *Blow-By-Effect* (will be explained in the failure analysis section). However, briefly, the notches are made to assure that a rapid energizing of the seal takes place at sudden changes in pressure and direction of motion.

Solution 2 for short-stroke and high-frequency: Reciprocating movement, with an increasing frequency above 5 Hz, the formation of lubrication under the contact face of the seal ring is reduced. High-frequency is most often occurring in connection with short-strokes. These two types of movements together accelerate the wear on hardware and seal. In this solution a symmetric seal is used with angled contact faces to ensure oil film is not scraped away from the surface. Oil is transported into the groove in the middle of the contact area forming an oil reservoir for lubrication. Wear particles are also likely to be captured in this groove, thus preventing them from embedding in the surface where the highest contact force occurs.

Solution 3 for better leakage control: The seal is designed with two seal edges. It acts as the primary seal for pressures from both sides, prevents build-up of hydrodynamic pressure over the seal profile, and prevents the risk of blow-by effect. The central sealing face increases the sealing effect.

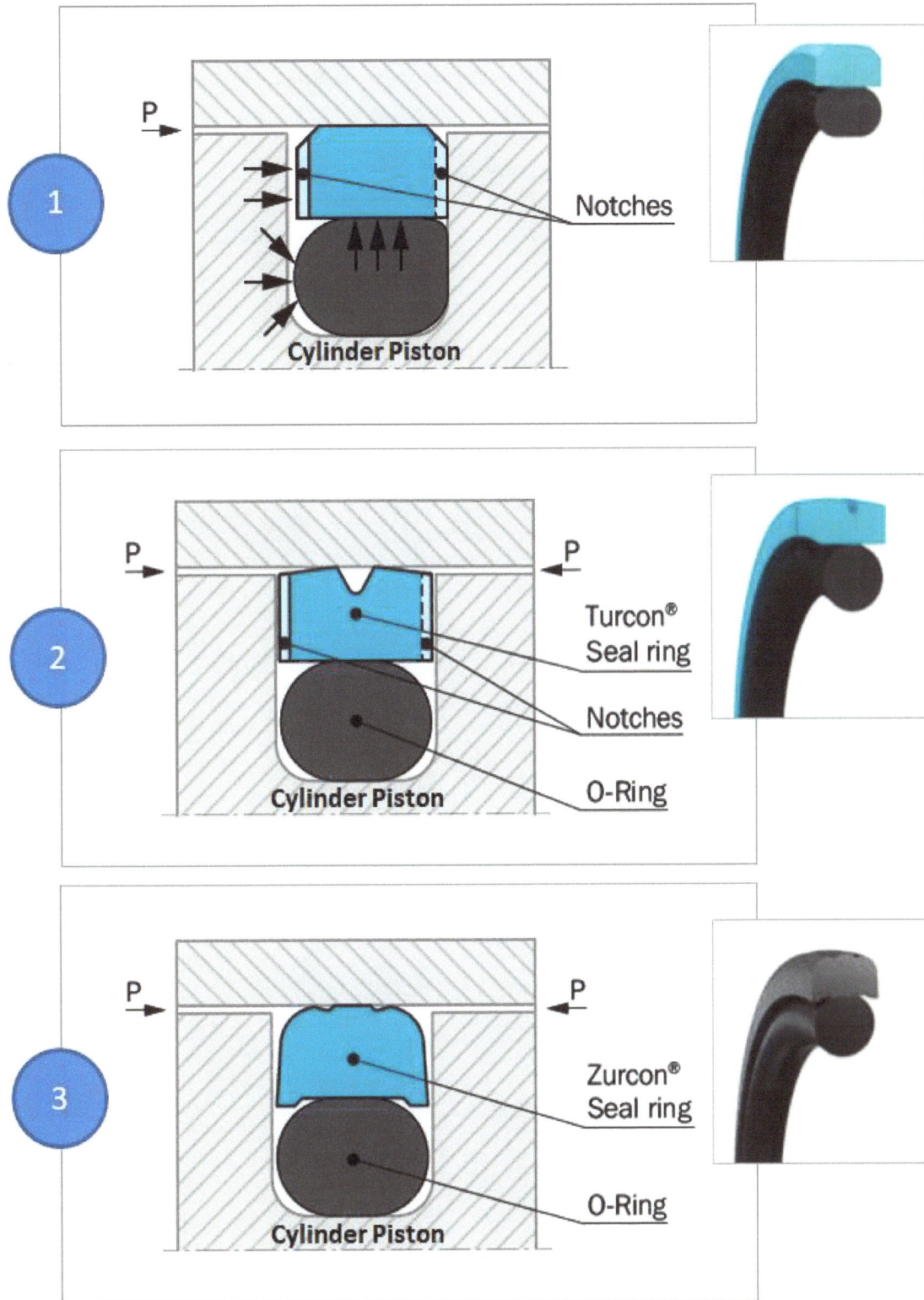

**Figure A.95 - Enhanced Bidirectional Piston Seal Designs for Medium-Duty Applications
(Courtesy of Trelleborg)**

Example 3: Bidirectional T-Shaped Piston Sealing for Heavy-Duty Applications
Figure A.96 shows a T-Shaped seal with two standard Backup-Rings. Its compact design provides improved stability and extrusion resistance in dynamic fluid sealing applications.

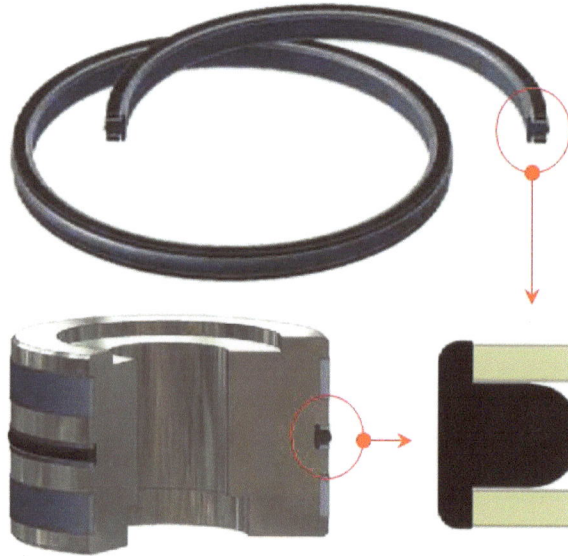

Fig. A.96 - Typical T-Shaped Seals for Bidirectional Piston Sealing (Courtesy of Parker)

This enhanced piston seal package, as shown in Fig. A.97, contains a heavy-duty squeeze-type T-Shaped piston seal and two L-Shaped Guide-Rings. This solution is good to seal across the piston during extension and retraction. The figure shows that the dynamic seal surface is designed with as wavy surface for better self-lubrication and reduced friction.

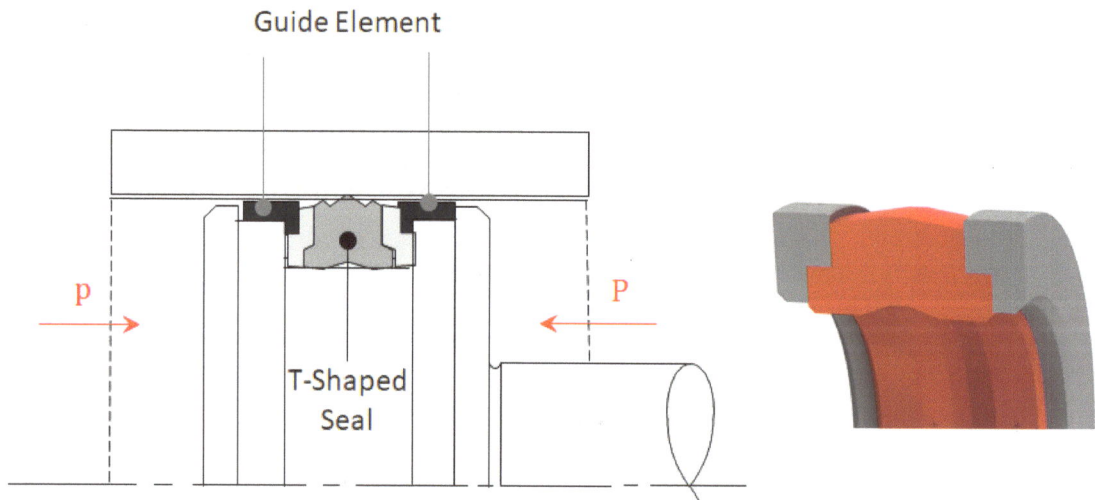

**Fig. A.97 - T-Shaped Seals for Bidirectional Piston Sealing
(Courtesy of American High-Performance Seals)**

Example 4: U-Cup Seals for Bidirectional Piston Sealing

Figure A.98 shows two single-acting U-cup seals are placed back-to-back to form bidirectional piston sealing solution. This piston sealing solution is good for cylinders that are pushing and pulling loads. The figure shows also a U-Cup seal for the cushioning head, a Guide-Ring, and an assembled magnet for cylinder position sensing.

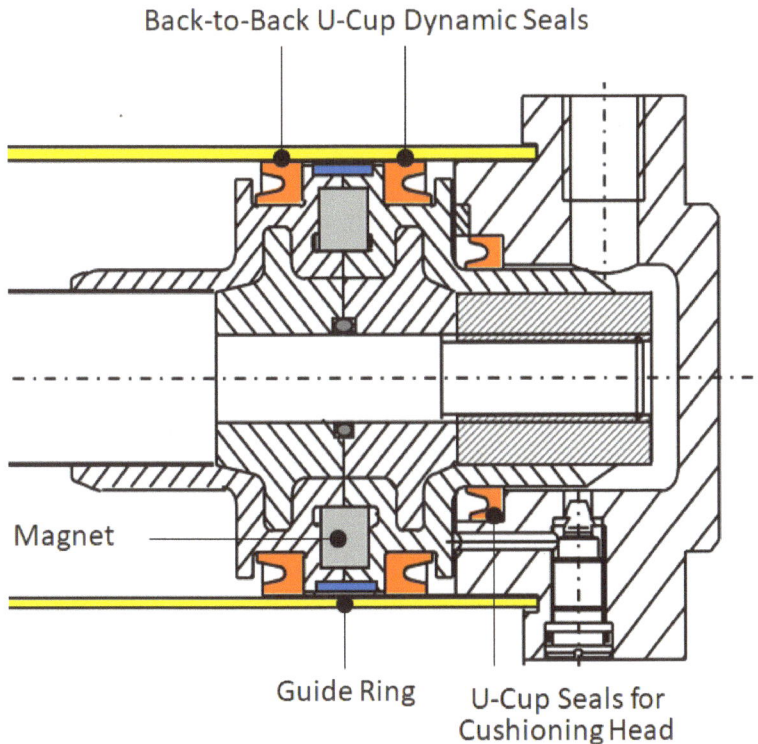

Fig. A.98 - U-Cup Seals for Bidirectional Piston Sealing

Figure A.99 shows a typical non-symmetrical hydraulic cylinder piston seal. Two seals can be placed on a piston, back to back, in separate glands offering bidirectional fluid sealing.

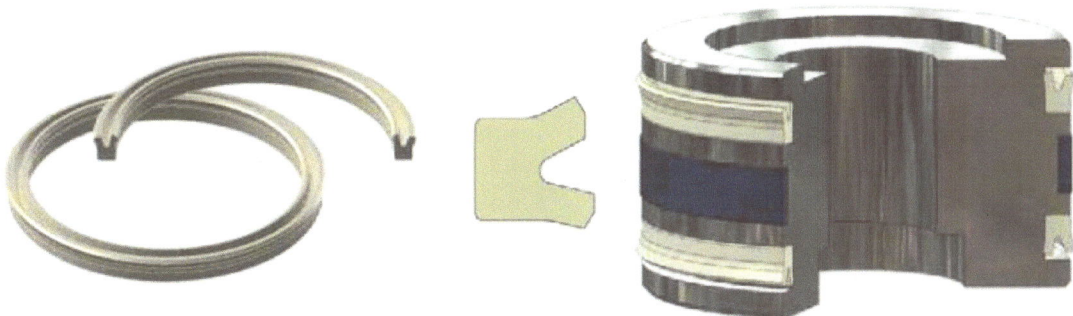

Fig. A.99 - Typical U-Cup Seals for Bidirectional Piston Sealing (Courtesy of Parker)

Figure A.100 shows a typical industry example of spring-energized U-Cup seals for bidirectional piston sealing.

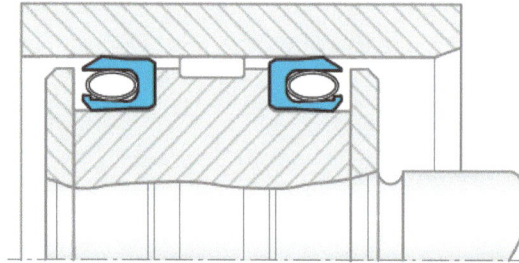

Fig. A.100 - Typical U-Cup Seals for Bidirectional Piston Sealing (Courtesy of Trelleborg)

7.4- Piston and Rod Design for Proper Sealing

There are some design issues that must be considered for proper sealing functions, otherwise the seal may be damaged during assembly and operation. Theses design issues are
- Design of *Seal Groove*.
- Design of cylinder piton and rod *Lead-in Chamfers*.
- Design of *Extrusion Gap*.
- Design of Mating *Surface Finish*.

7.4.1- Design of Seal Groove and Lead-In Chamfers

When designing a seal groove, sharp edges must be eliminated by proper rounding. Figure A.101 shows the effect of improper lead-in chamfers on an O-Ring. During installation, the improper lead-in chamfer on cylinder piton and rod cause tearing of the O-Ring during installation. Seals manufacturers also provide instructions for lead-in chamfers based on seal size.

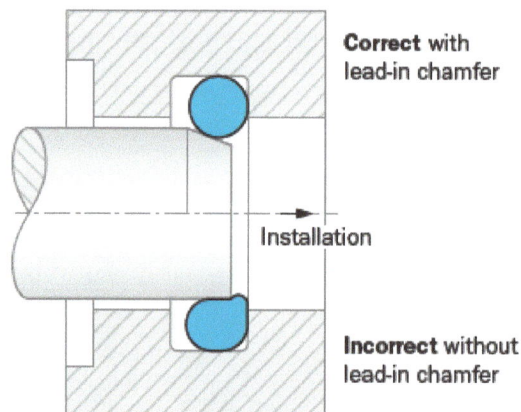

Correct with
lead-in chamfer

Installation

Incorrect without
lead-in chamfer

**Fig. A.101 - Improper Lead-in Chamfer Cause Tearing of an O-Ring During Installation
(Courtesy of Trelleborg)**

Therefore, seal manufacturers must provide design instructions based on the cylinder size and working conditions. Figure A.102 provide an industry example for a double-acting piston seal.

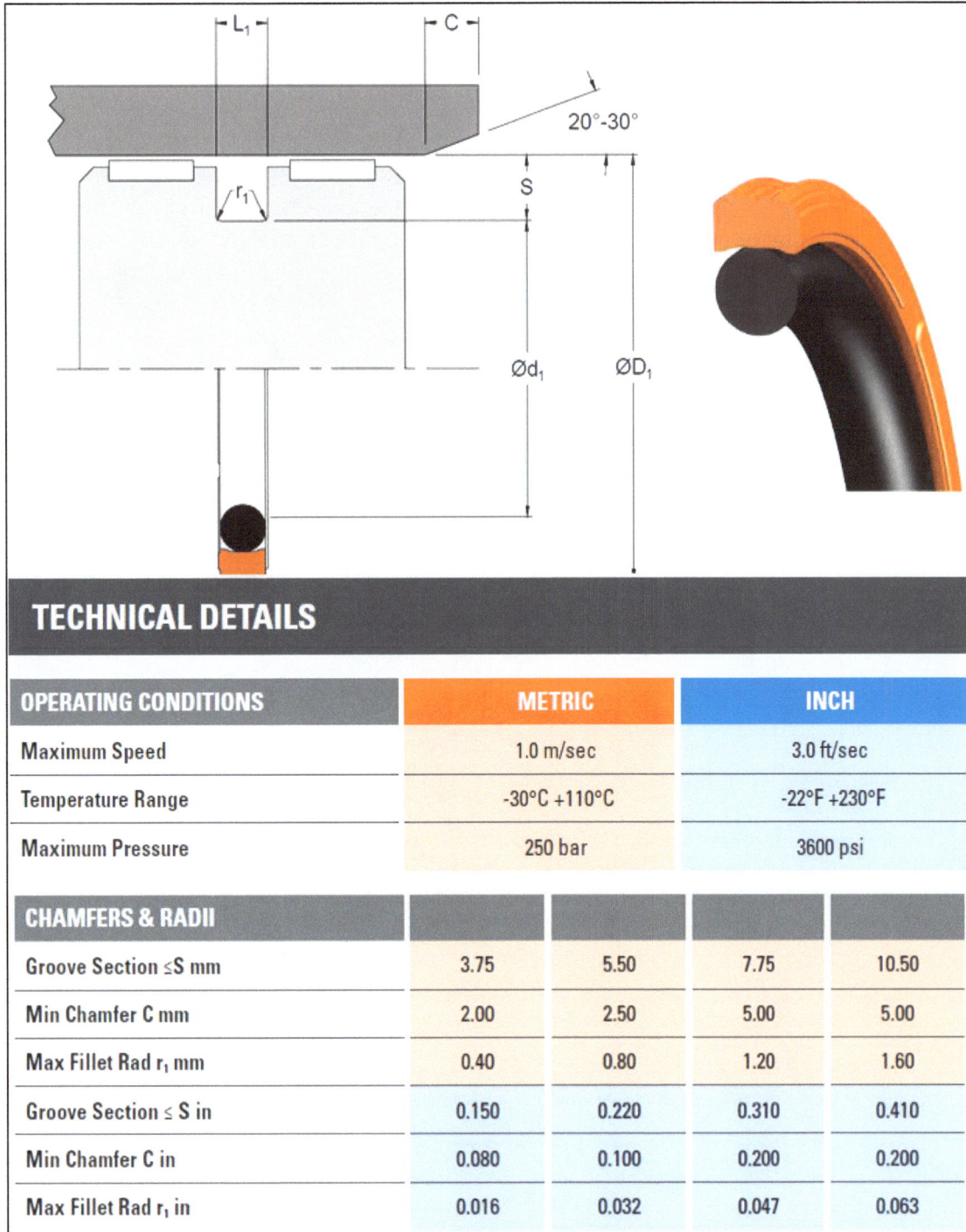

TECHNICAL DETAILS

OPERATING CONDITIONS	METRIC	INCH
Maximum Speed	1.0 m/sec	3.0 ft/sec
Temperature Range	-30°C +110°C	-22°F +230°F
Maximum Pressure	250 bar	3600 psi

CHAMFERS & RADII				
Groove Section ≤S mm	3.75	5.50	7.75	10.50
Min Chamfer C mm	2.00	2.50	5.00	5.00
Max Fillet Rad r_1 mm	0.40	0.80	1.20	1.60
Groove Section ≤ S in	0.150	0.220	0.310	0.410
Min Chamfer C in	0.080	0.100	0.200	0.200
Max Fillet Rad r_1 in	0.016	0.032	0.047	0.063

Fig. A.102 - Data Sheet for a Double Acting Piston Seal 764 with Pre-Loaded O-Ring (Courtesy of Hallite Seals)

7.4.2- Extrusion Gap Design

Seal manufacturers provide guidelines for designing the extrusion gap.

Example 1: Use of graphs provided by the manufacturer:
Figure 4.103 shows that the *Extrusion Limits* are defined based on the seal hardness and working pressure. Softer seals extrude at low pressure. Therefore, at same extrusion gap (diametrical clearance), the required seal hardness is proportional to working pressure. Softer seals require less extrusion gap. Therefore, for same working pressure, the allowable extrusion gap is proportional to the seal hardness. The figure also shows that, at the same working pressure, the required extrusion gab is inversely proportional to seal hardness. The figure states to reduce the clearance shown by 60% when Silicon or Fluro silicone are used because the show less resistance to extrusion.

Case Study: For fluid pressure of 55.2 bar (800 psi) and a seal hardness 70 Shore A, maximum allowable gap extrusion is 0.3 mm (0.01 in)

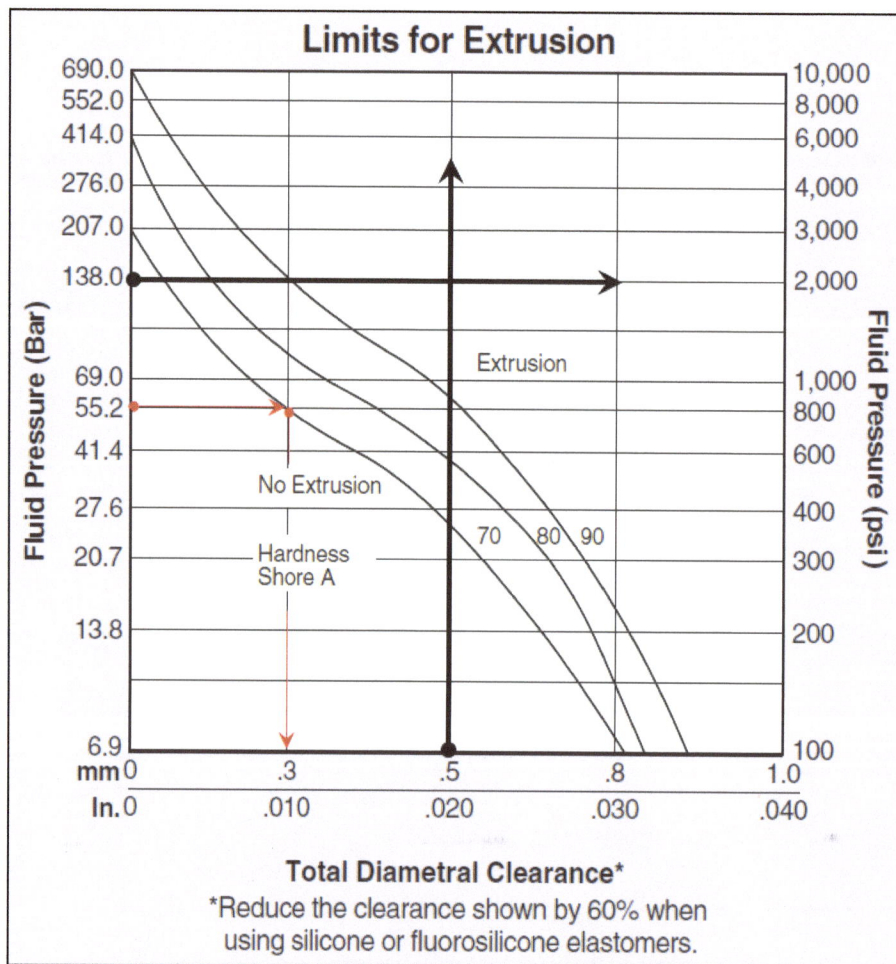

Fig. A.103 - Limits of Extrusion (Courtesy of Parker)

Example 2- Use of dimensional parameters and a nomograph provided by the manufacturer

Figure A.104 shows the following dimensional parameters:

e = Maximum sealing and anti-extrusion gap.
D = Piston diameter.
d = Rod diameter.
S = Cross section.
P = Working pressure
T = Working Temperature.

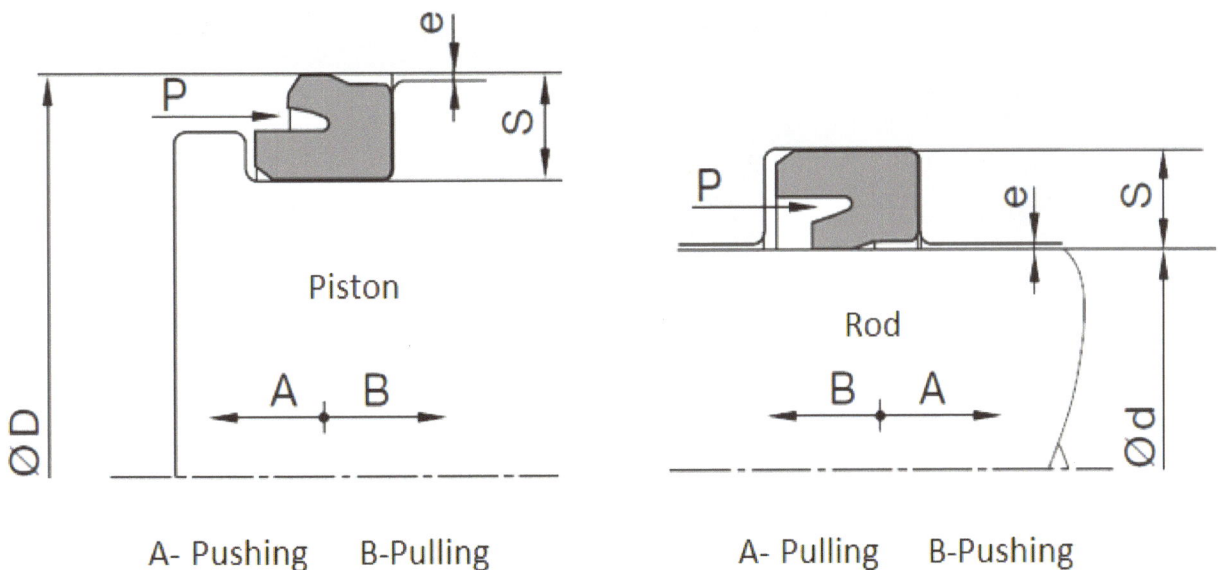

A- Pushing   B-Pulling   A- Pulling   B-Pushing

Fig. A.104 - Anti-Extrusion Gap (Courtesy of Parker)

The nomograph, presented in Fig. A.105, is developed for finding the maximum allowable sealing gap. The chart is based on the worst-case scenario such as the pushing case and the softest material. This chart is applicable for seals with 70-85 Shore A hardness.

Method of using the chart is as follows:
1- Draw the line connecting **d/D** to **S** and extend it until it interests with the line $\xi 1$.
2- Draw the line connecting **P** to **T** and extend it until it interests with the line $\xi 2$.
3- Connect the two intersections and read the allowable sealing gap.

Case Study: for d/D = 100 mm, S = 6 mm, P = 100 bar, and T = 80 OC, the sealing gap e = 0.18 mm

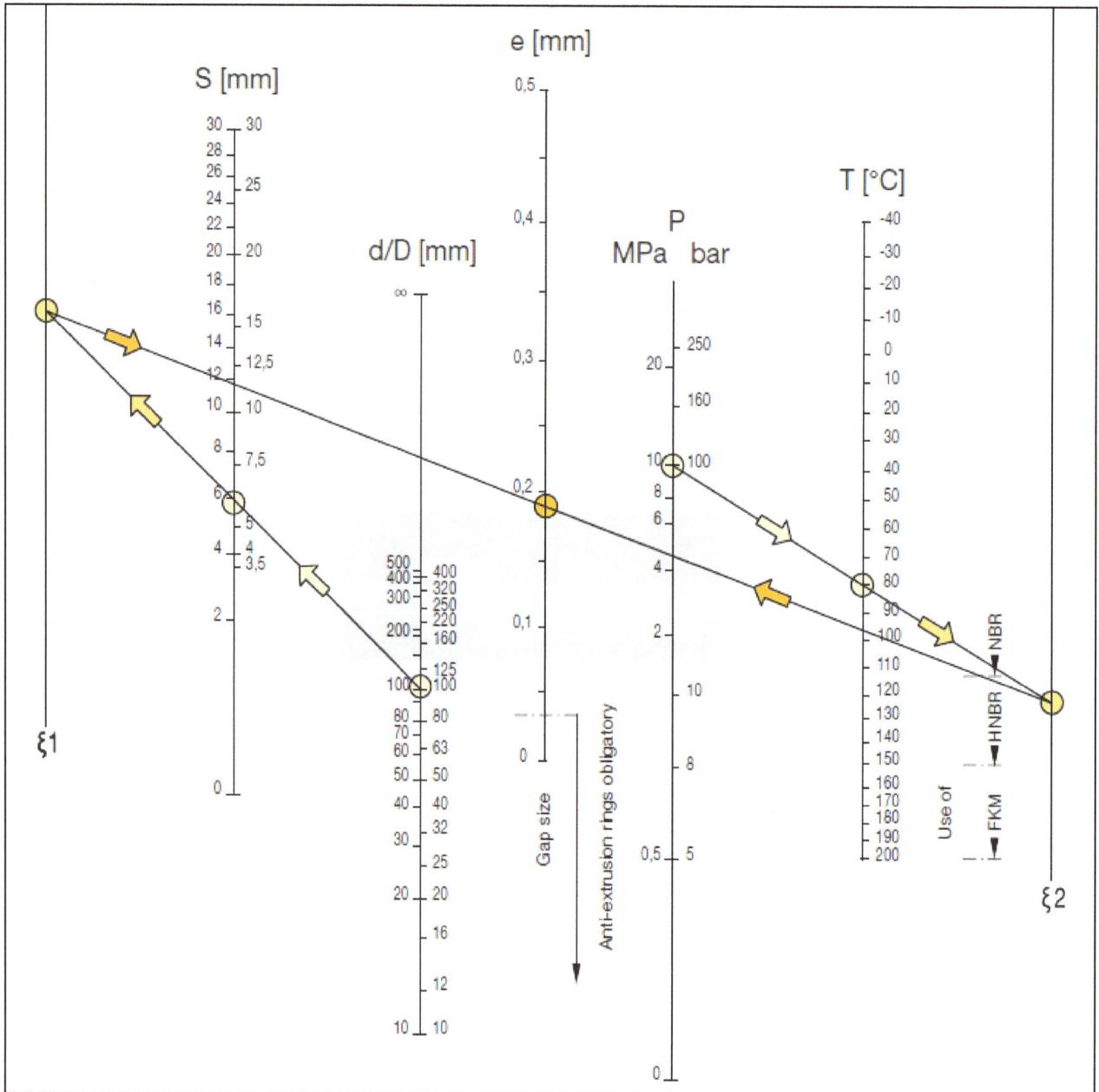

Fig. A.105 - Anti-Extrusion Gap Nomograph (Courtesy of Parker)

Example 3- Use of tabulated results provided by the manufacturer:

Table A.10 shows an industry example.

MAXIMUM EXTRUSION GAP			
Pressure bar	100	160	250
Maximum Gap mm	0.60	0.50	0.40
Pressure psi	1500	2400	3750
Maximum Gap in	0.024	0.020	0.016

Table A.10 - Data Sheet for a Double Acting Piston Seal 764 with Pre-Loaded O-Ring (Courtesy of Hallite Seals)

7.4.3- Design of Mating Surface Finish

Figure A.106 shows the dynamic and static sealing surfaces in a hydraulic cylinder. Proper finish of mating surfaces in contact with the seals is critical in assuring maximum seal performance and service life within a given application. If the dynamic sealing surface is too smooth, it will improperly retain lubrication film. Poor seal lubrication causes excessive seal wear due to frictional heat. If the dynamic sealing surface is too coarse, premature seal failure may occur due to the roughness of the surface, hence causing small cuts or scores in the sealing lip. The static sealing surface finish must not be ignored in the control of leakage. Generally, static sealing surfaces should be free from chatter marks.

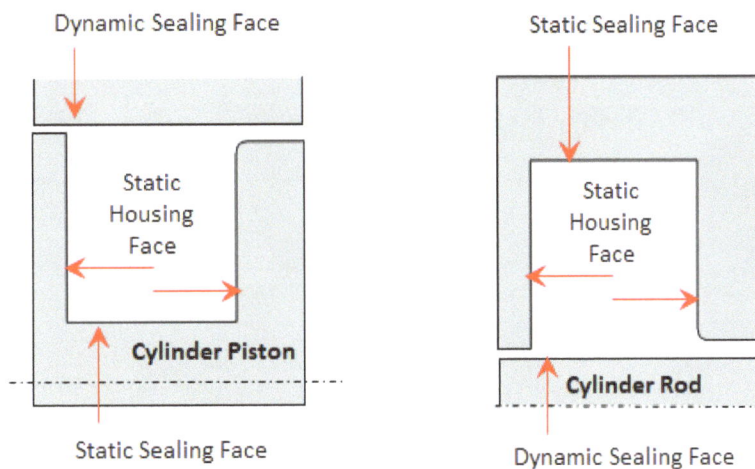

Fig. A.106 - Dynamic and Static Sealing Surfaces in a Hydraulic Cylinder

Many parameters can be used to define surface finishes, which are explained in *ISO 4287* and *ISO 4288*. The following are the most commonly used surface finish measurements in the fluid power industry.

Figure A.107 shows the surface finish measurement **Ra** is the arithmetical mean deviation of an absolute ordinate over the evaluation length. Figure A.108 shows the surface finish measurement **Rt** is the Sum of height of the largest profile peak height R_p and the largest profile valley R_v over the evaluation length. Figure A.109 shows the surface finish measurement **Rz(n)** is the Sum of height of the largest profile peak height R_p and the largest profile valley R_v within a sampling length. the surface finish measurement **Rz** is then the average of **$R_{z(n)}$** over the evaluated length. **Ra** and **Rt** are the most common measurement types associated with surface finish.

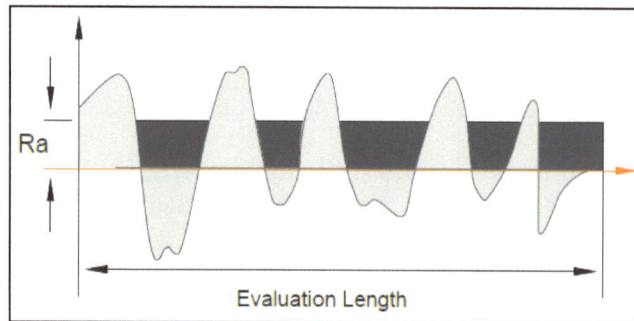

Fig. A.107 - Surface Finish Measurement Ra (Courtesy of Hallite Seals)

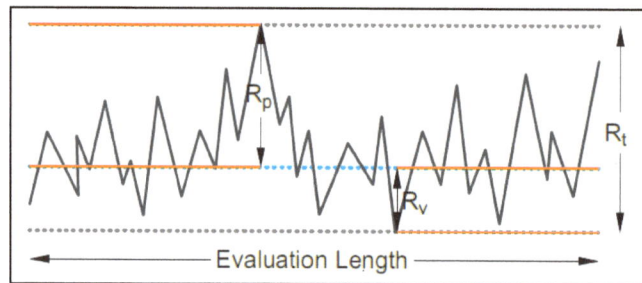

Fig. A.108 - Surface Finish Measurement Rt (Courtesy of Hallite Seals)

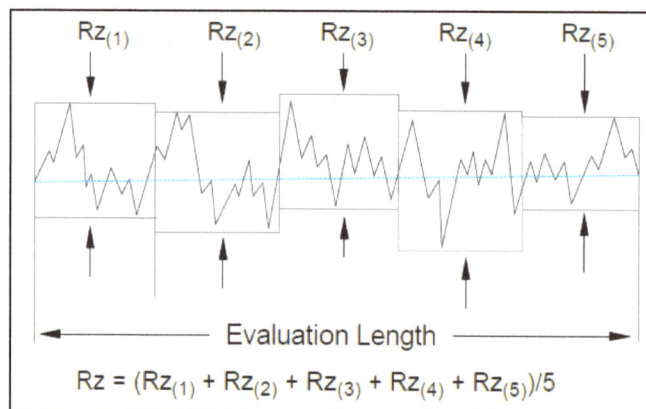

Fig. A.109 - Surface Finish Measurement R_Z (Courtesy of Hallite Seals)

Table A.11 shows an industry example.

SURFACE ROUGHNESS	µmRa	µmRz	µmRt	µinRa	µinRz	µinRt
Dynamic Sealing Face ØD₁	0.1 - 0.4	1.6 max	4 max	4 - 16	63 max	157 max
Static Sealing Face Ød₁	1.6 max	6.3 max	10 max	63 max	250 max	394 max
Static Housing Faces L₁	3.2 max	10 max	16 max	125 max	394 max	630 max

Table A.11 - Data Sheet for a Double Acting Piston Seal 764 with Pre-Loaded O-Ring (Courtesy of Hallite Seals)

Chapter 8 – Sealing Solutions for Rotational Shafts

Rotational Dynamic Seals are used to prevent external leakage along rotating shafts in hydraulic pumps, motors, and rotary actuators. They are designed for axial, radial, or combined sealing directions. Like hydraulic seals for linear shafts, as shown in Fig. A.110, materials and designs for a rotational shaft seals vary depending on working conditions and applications. The following subtitles presents most commonly used hydraulic sealing solutions for rotational shafts.

Fig. A.110 - Various Materials and Designs for Rotational Shaft Seals (Courtesy of American High-Performance Seals)

8.1- Rotational Radial Seals

A *Rotational Radial Seal* acts radially against the rotating shaft to prevent axial leakage. Figure A.111 shows the most common rotational radial seal. Figure A.112 shows the construction details of the rotational radial seal shown in the previous figure. As shown in the figure, such a seal holds the rotating shaft radially preventing axial leakage.

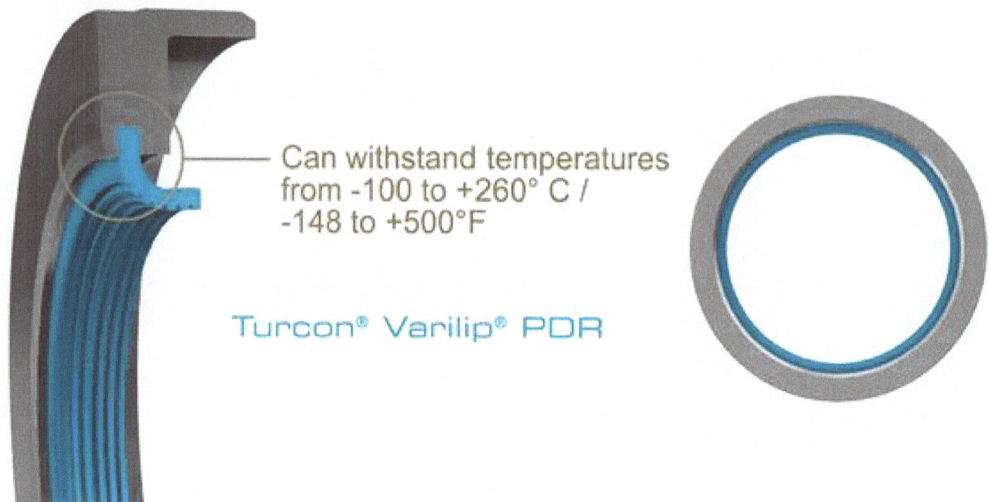

Can withstand temperatures
from -100 to +260° C /
-148 to +500°F

Turcon® Varilip® PDR

Fig. A.111 - Most Common Rotational Radial Seal

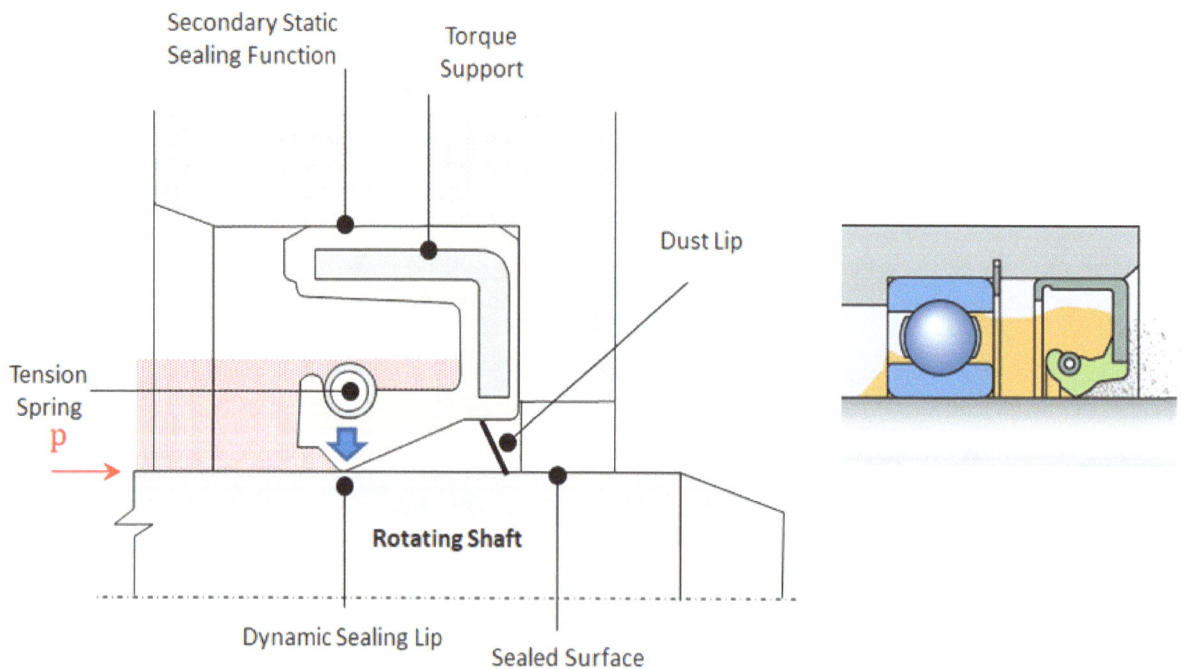

Fig. A.112 - Heavy Duty Rotational Radial Seals

Figure A.113 shows typical rotational radial seals. They are available for wide range of pressure anywhere from 1 to 700 bar (15 to 10,152 psi). These seals are comprised of standard and special seal designs of rubber, with or without metal supporting inserts, with or without a spring, and non-metal rotary shaft lip seals. The dynamic sealing lips are made of rubber or special elastomers to meet specific fluid, temperature and other demanding operating requirements. Typical applications: oil and grease retention for power transmissions, motors, pumps, gearboxes, fans, and machine tools.

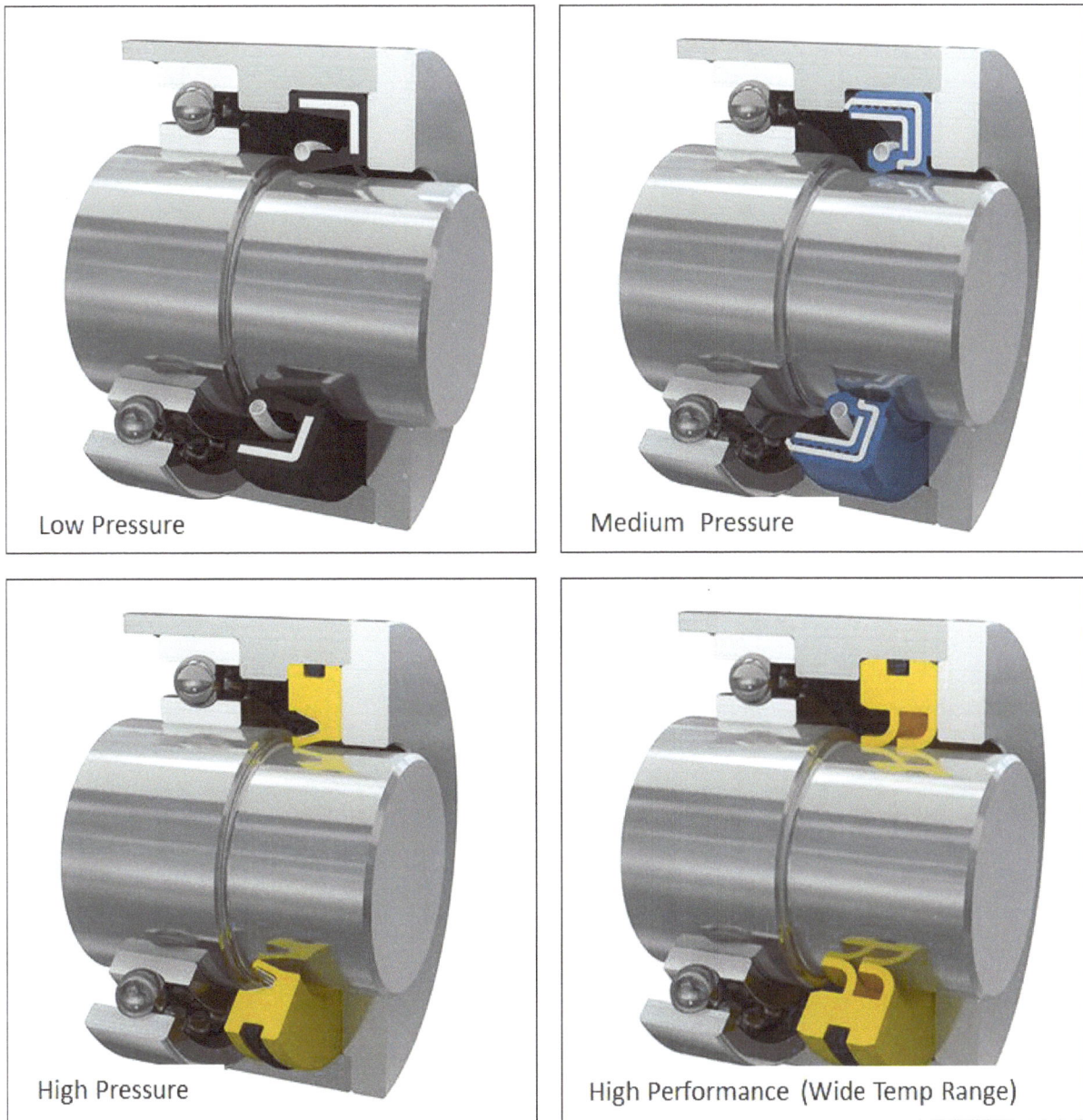

Low Pressure

Medium Pressure

High Pressure

High Performance (Wide Temp Range)

**Fig. A.113 - Construction Details of the Most Common Rotational Radial Seals
(Courtesy of American High-Performance Seals)**

Figure A.114 shows standard configurations of rotational radial seals. Waved rubber outer surface improves the retaining force between the seal and the groove in the housing, hence preventing rotation of the seal in its groove.

With Rubber Case		With Metallic Case	
Flat Surface	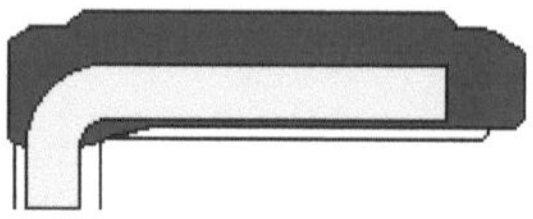	DIN Type B without Dust Lip without Support	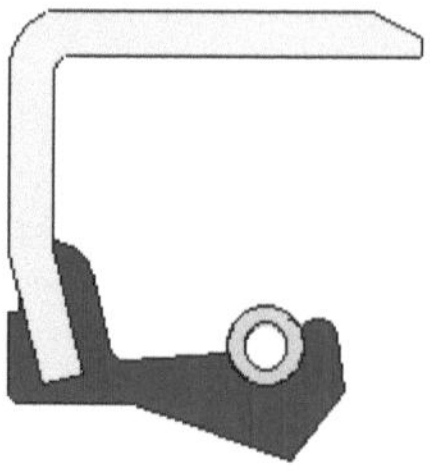
DIN Type A Waved Surface without Dust Lip	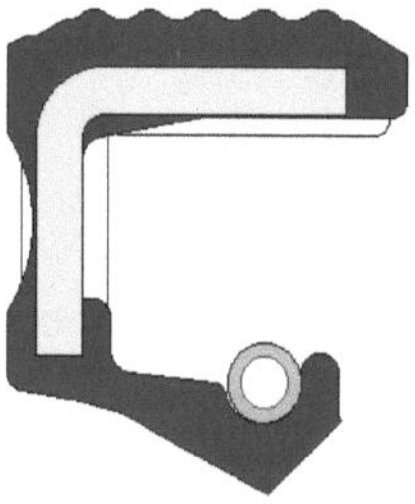	DIN Type BS with Dust Lip without Support	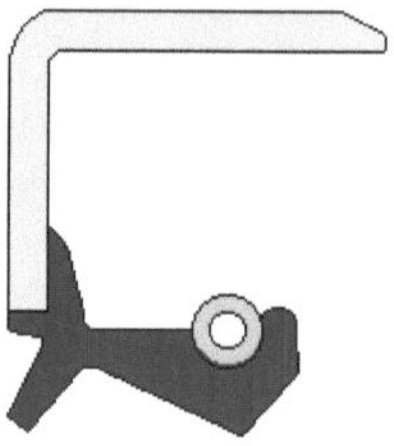
DIN Type AS Waved Surface with Dust Lip	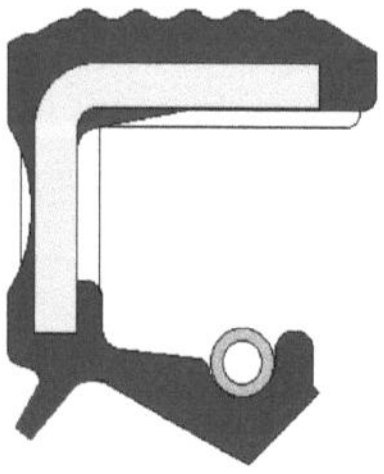	DIN Type C without Dust Lip with Support	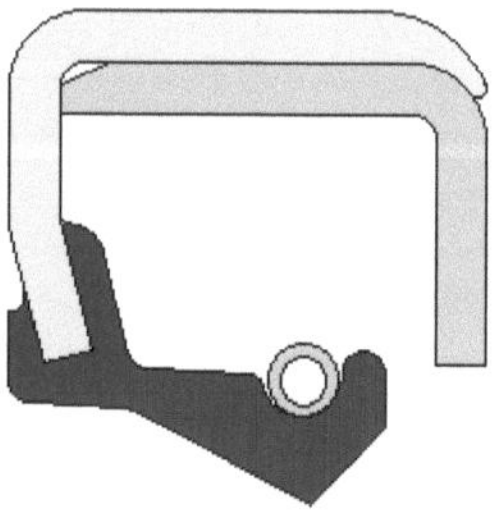
		DIN Type CS with Dust Lip with Support	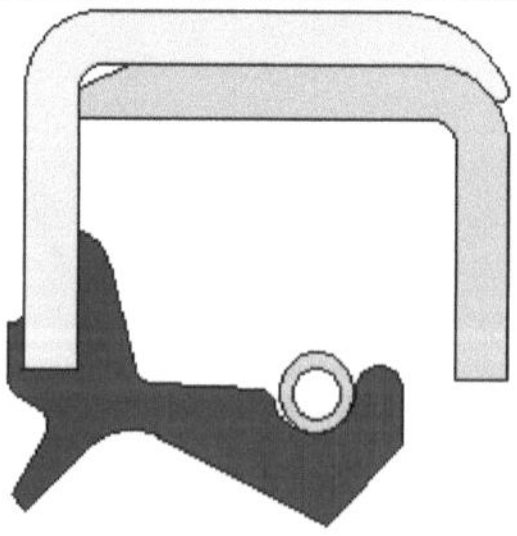

Fig. A.114 - Configurations of Rotational Radial Seals to DIN 3760/ISO 6194 Standards

8.2- Rotational Axial Seals

A *Rotational Axial Seal* acts axially against the shaft to prevent radial leakage. The following are common designs for rotational axial seals.

V-Ring Rotational Axial Seals: Figure A.115, The *V-RING* axial seal is a unique all-rubber seal for rotary shafts. Such a seal is used for a broad range of applications. It can also be used as a secondary seal to protect a primary seal that do not perform well in hostile environments. This seal is normally stretched and mounted directly on the shaft, where it is held on the shaft by the inherent tension of the rubber body. The sealing lip is flexible and applies only a relatively light contact pressure against the counter-face and yet is still sufficient to maintain the sealing function.

As shown in Fig. A.116, together with a *Clamping Band,* a V-Ring axial seal can be used for shaft diameters larger than 1,500 mm.

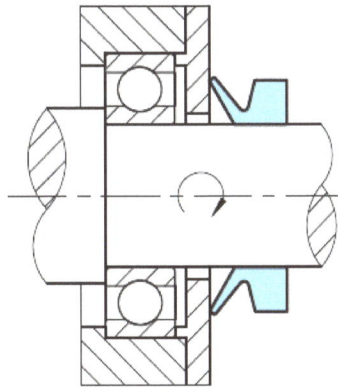

Fig. A.115 - Classic Rotational Axial Seal (Courtesy of Trelleborg)

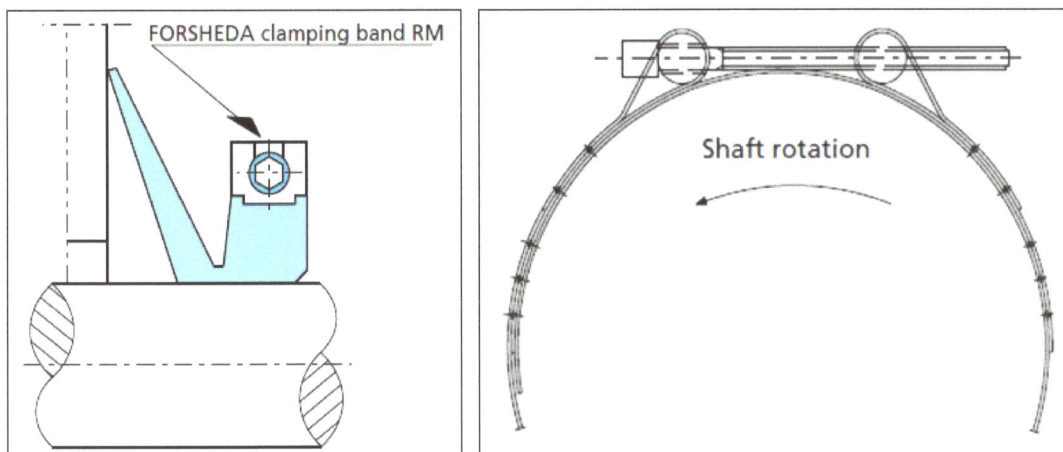

Fig. A.116 - Classic Rotational Axial Seal (Courtesy of Trelleborg)

Power Losses in Rotational Shafts: A test was made for two types of V-Rings in contact with unhardened steel surface in dry-running condition. Figure 4.117 shows the power losses versus the peripheral (tangential) speed for various shaft diameters (d in mm). Once breakaway friction is overcome, the friction increases steadily until around the 12 m/s range, when it reduces quite quickly.

Due to the centrifugal force, the contact pressure of the lip decreases with increased speed. This means that frictional losses and heat are kept to a minimum, resulting in excellent wear characteristics and extended seal life. In the 15 - 20 m/s range the friction reduces to zero. The V-Ring then serves as a clearance seal and deflector.

Generally speaking, the power losses resulting from a V-Ring are always lower than a corresponding radial oil seal.

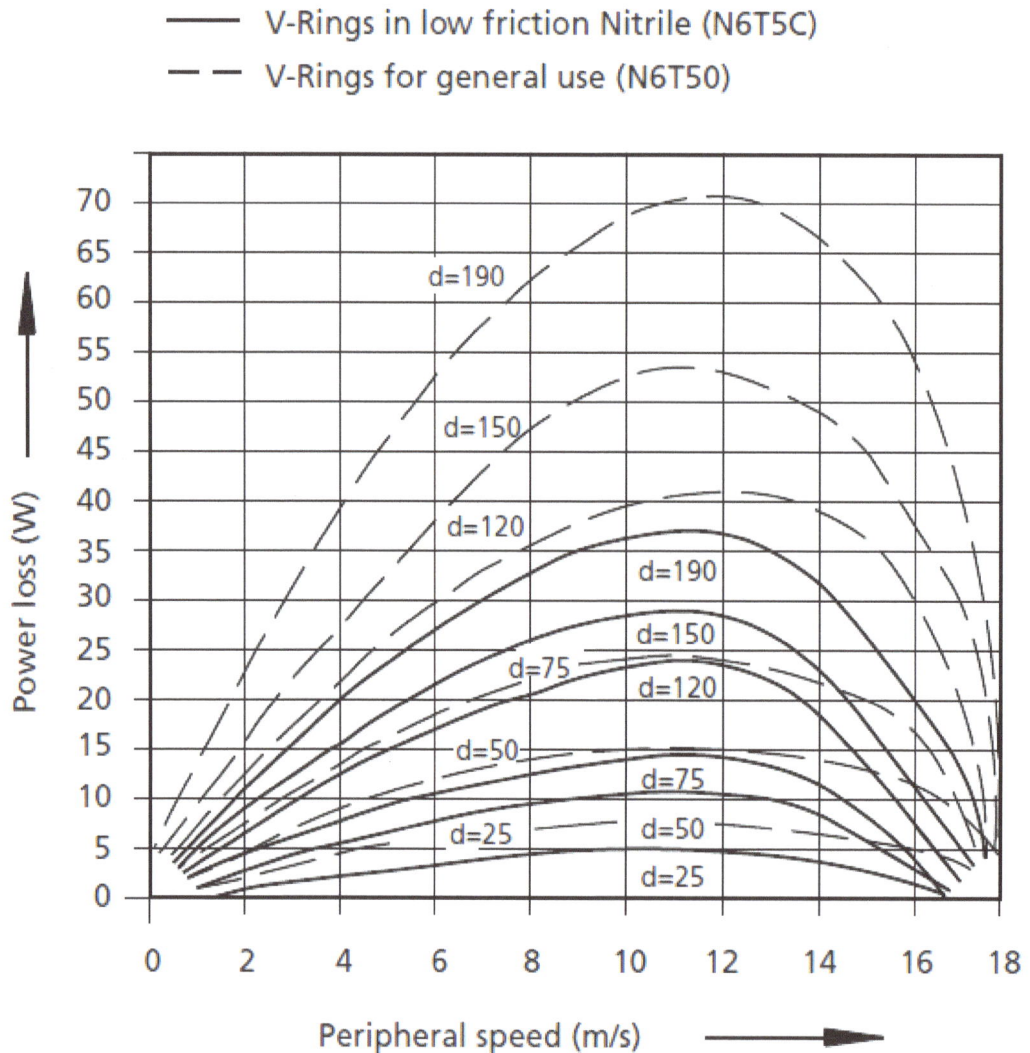

Fig. A.117 - Power Losses versus Peripheral Speed (Courtesy of Trelleborg)

GAMMA Rotational Axial Seals: Figure A.118 shows an enhanced rotational axial seal named *GAMMA Seal*. It consists of two parts, sealing element and metal case. The GAMMA seal is designed to be fixed to the rotational shaft at a predetermined distance from the sealing surface. The GAMMA seal is primarily intended for sealing against foreign matter, liquid splatter, and grease.

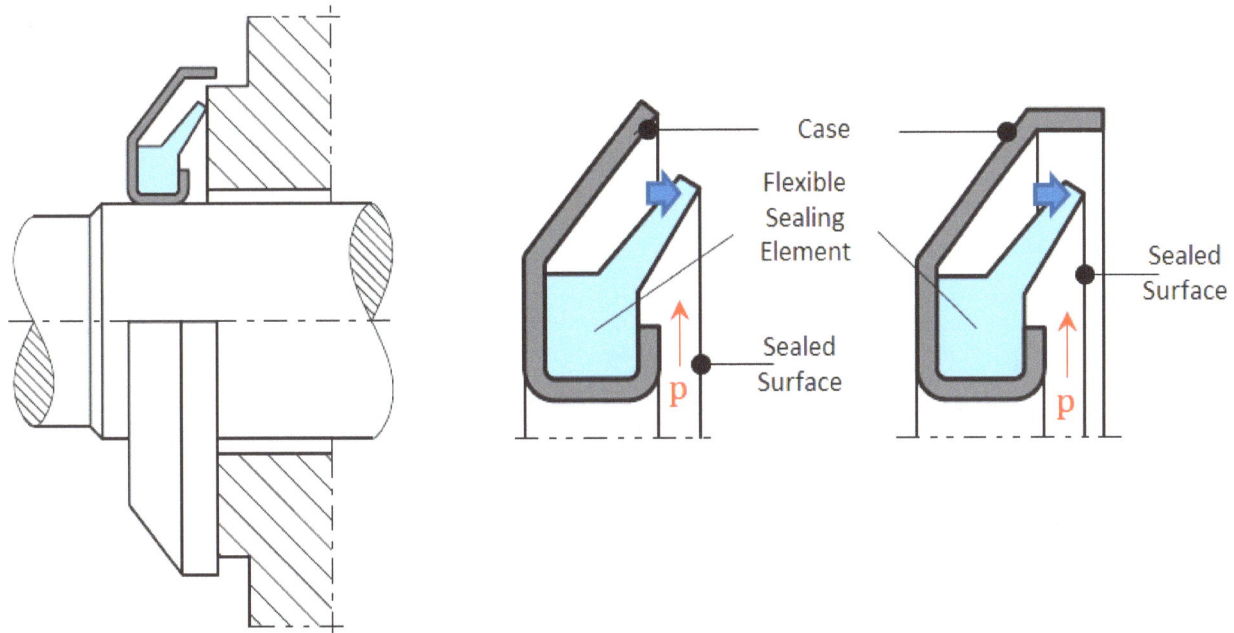

Fig. A.118 - GAMMA Rotational Axial Seals (Courtesy of Trelleborg)

Enhanced Axial Shaft Seals: As shown in Fig. A.119, axial shaft seals are used primarily as a protective seal for bearings. Their sizes are matched to those of roller bearings. If fluids are to be prevented from escaping, a design with an internal seal lip, is preferred. The design with external sealing lip is suitable for sealing grease and for protection against dirt entering from the outside.

Both types consist of an elastomer-elastic membrane with a metallic reinforcement ring. The membrane has an axial sealing lip. The elastomeric sealing lip is axially spring-loaded against the opposite mating face by a spider spring. The sealing lip is designed in a conical form to obtain a minimum contact area, thus considerably reducing friction, heat and wear.

Internal Sealing   External Sealing

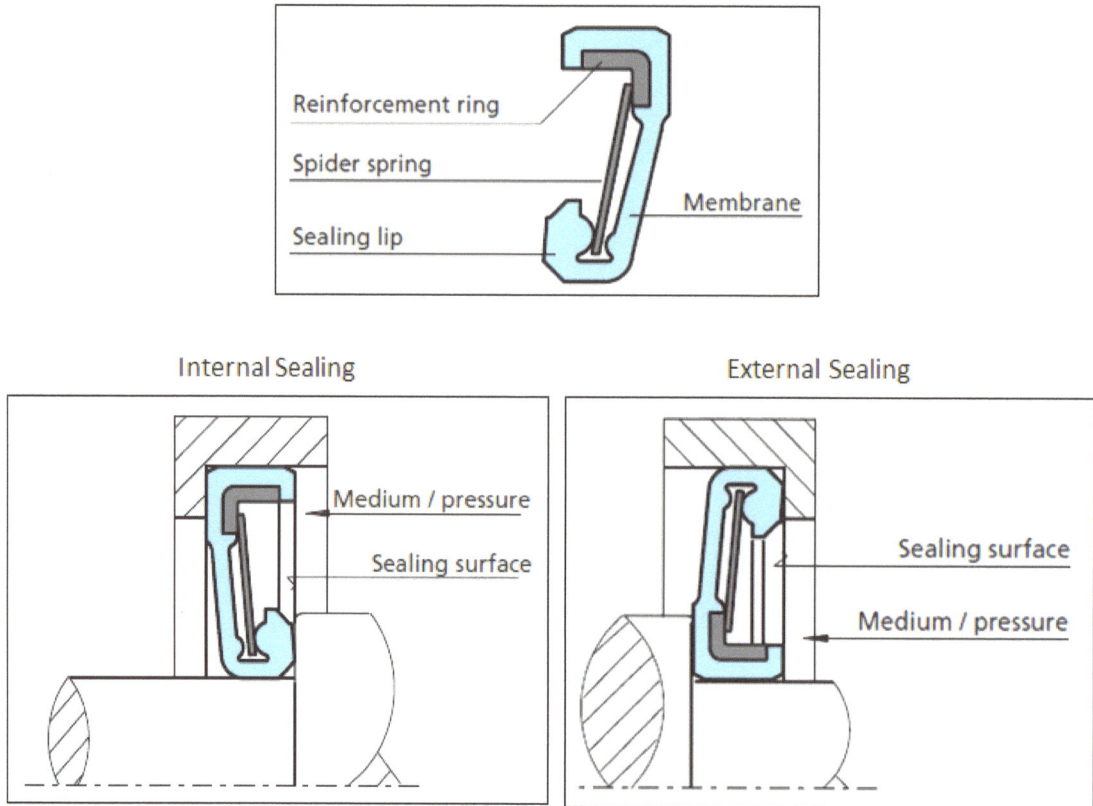

Fig. A.119 - Enhance Rotational Axial Seals (Courtesy of Trelleborg)

8.3- Combined Axial/Radial Sealing Solutions for Rotational Shafts

As shown in Fig. A.120, a flexible rubber V-Ring rotational axial seal is stretch-fit onto the shaft and rotate with the shaft against a counter face. It is used as a primary seal to protect the shaft against contaminants. The spring-energized rotational radial seal is a secondary seal to prevent external leakage.

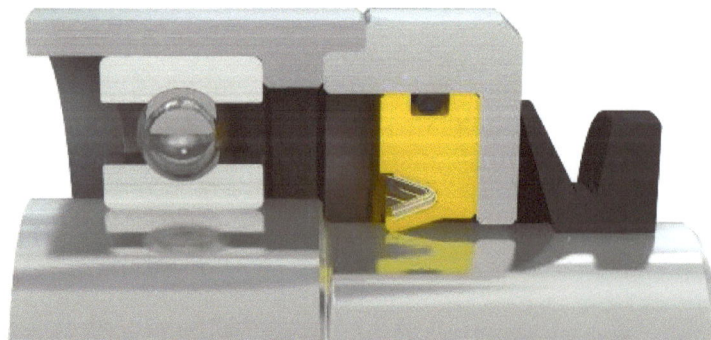

**Fig. A.120 - Combined Axial/Radial Sealing Solutions for Rotational Shafts
(Courtesy of American High-Performance Seals)**

Chapter 9 – Best Practices for Hydraulic Seals Installation

Proper installation of a hydraulic seal is an important process for the seal to perform reliably. Basically, the first advice is to review the seal installation instructions provided by the manufacturer. If not found, the following list provides guidelines for installing hydraulic seals.

Best Practices for Hydraulic Seals Installation:
1. **Figure A.121, Properly Remove the Old Seal:** Carefully dismount the old seal without damaging the bores or shafts.
2. **Figure A.122, Use Genuine Seals:** DO NOT use non-branded seals or seals that are not approved by the hydraulic component manufacturer.
3. **Figure A.123 Never use a Pretensioned or Pre-Used Seal:** That is because used seals may be plastically deformed and have defects or geometrical shape changes that may not be seen by your naked eyes. In most cases where components must be disassembled for inspection, seals should be replaced.
4. **Figure A.124, Inspect New Seals:** DO NOT use sharp objects to remove a seal out of its package. Before mounting the new seal, inspect it for damage on the circumference of the sealing lip or the outer diameter. DO NOT shorten the original tension spring if found. Double check the correct placement direction of the seal.
5. **Figure A.125, Inspect Seal Groove:** It should be clean and free of damage or sharp edges.
6. **Figure A.126, Inspect Assembly Tools:** They should be routinely inspected and calibrated. Before using them, make sure they are clean and free of sharp edges or scratches.
7. **Figure A.127, Cover Threads:** If the seal must be stretched over sharp edges, such as threaded parts or lead-in chamfers, these areas must be covered to prevent damaging the seals during the assembly process.
8. **Figure A.128, Lubricate Seal Before Installation:** Lubricate the seal and the mounting surface before seal installation. That reduces the surface friction on the seal and makes it easy to install it. Do NOT use regular grease. Use the same hydraulic fluid or a predefined compatible seal lubricant.
9. **Figures A.129 and A.130, Use Adequate Installation Tools:** Use adequate tools for installing rod and piston seals to prevent damaging or twisting the seals.
10. **Figure A.131, Compress the Seal in kidney Shape:** If needed or no assembly tools are found, compress the seal into a kidney shape. If the seal has a notch, DO NOT bend it from the position of the notch as this may cause overstretch or damage to the seal material.
11. **Figure A.132, Check Proper Installation of the Seal:** make sure the seal is not tilted or twisted, and the seal axis coincids with the shaft or piston axis.
12. **Figure A.133, Squeeze the Seal in its Groove:** In both static and dynamic applications, a certain amount of squeeze or compression is required upon installation to maintain contact with the sealing surfaces and prevent fluid leakage. This can be done by installation cones.

Fig. A.121 - Properly Remove the Old Seal

Fig. A.122 - Use Genuine Seals (Courtesy of Parker)

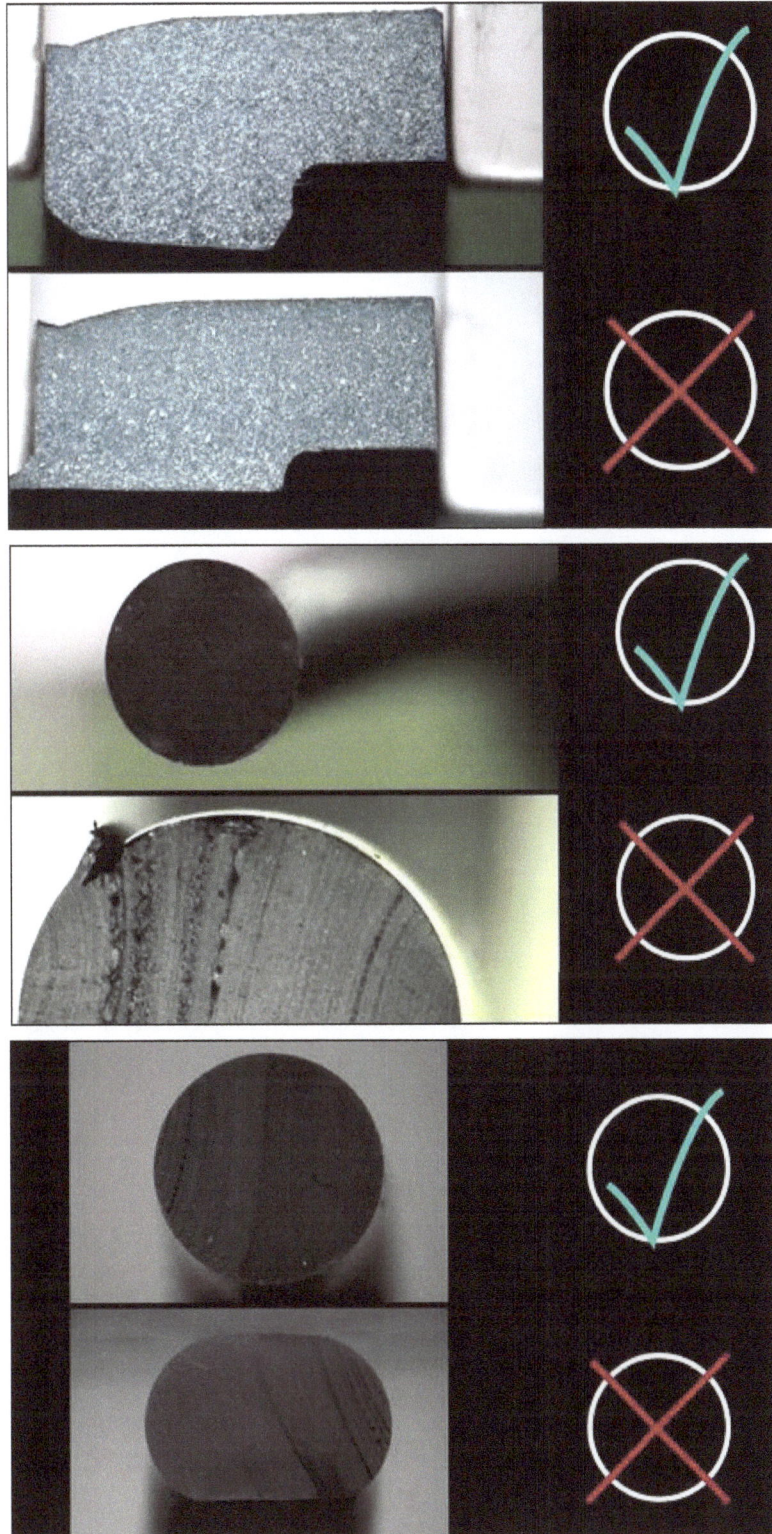

Fig. A.123 - Never use a Pretensioned or Pre-Used Seal (Courtesy of Trelleborg)

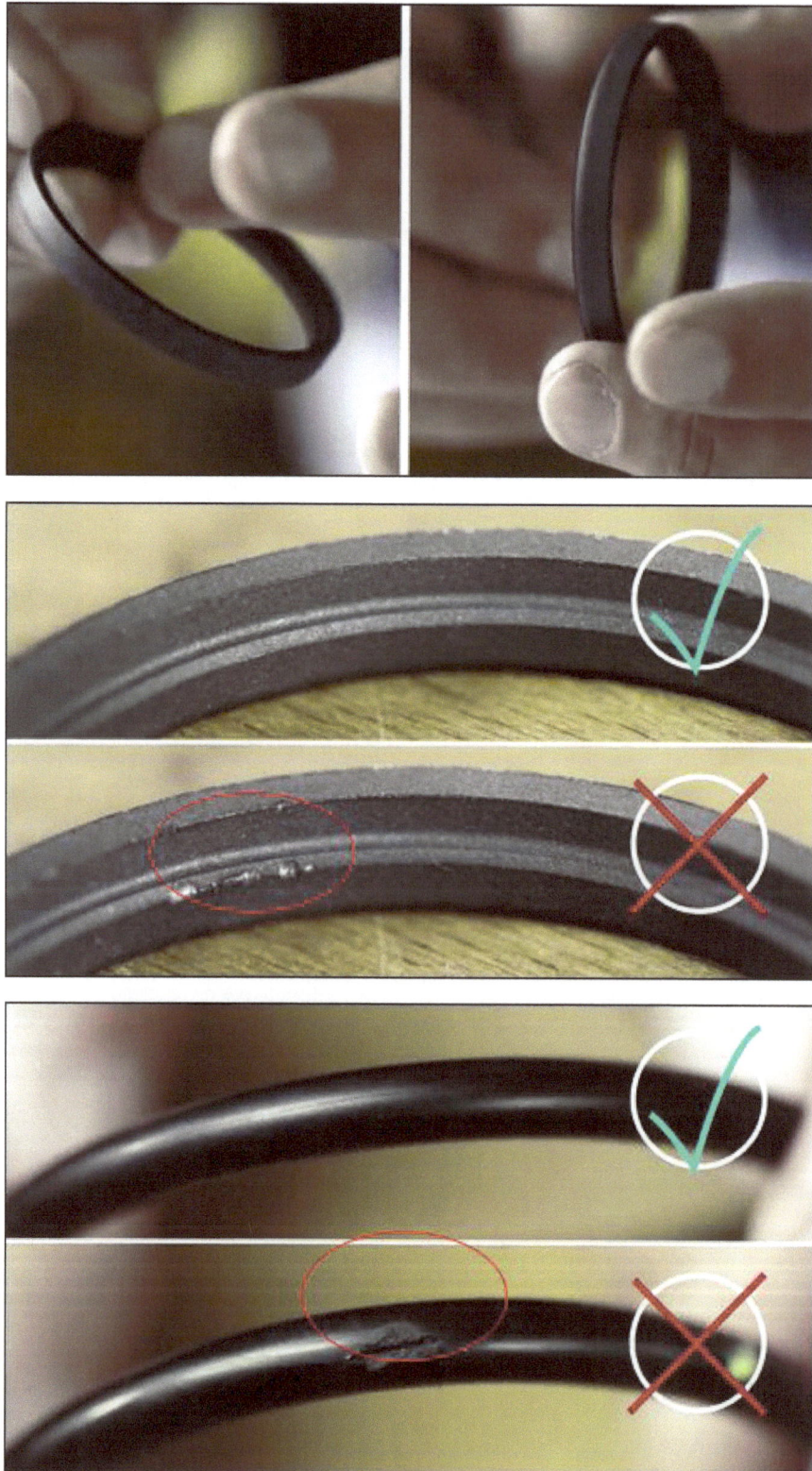

Fig. A.124 - Inspect New Seals (Courtesy of Trelleborg)

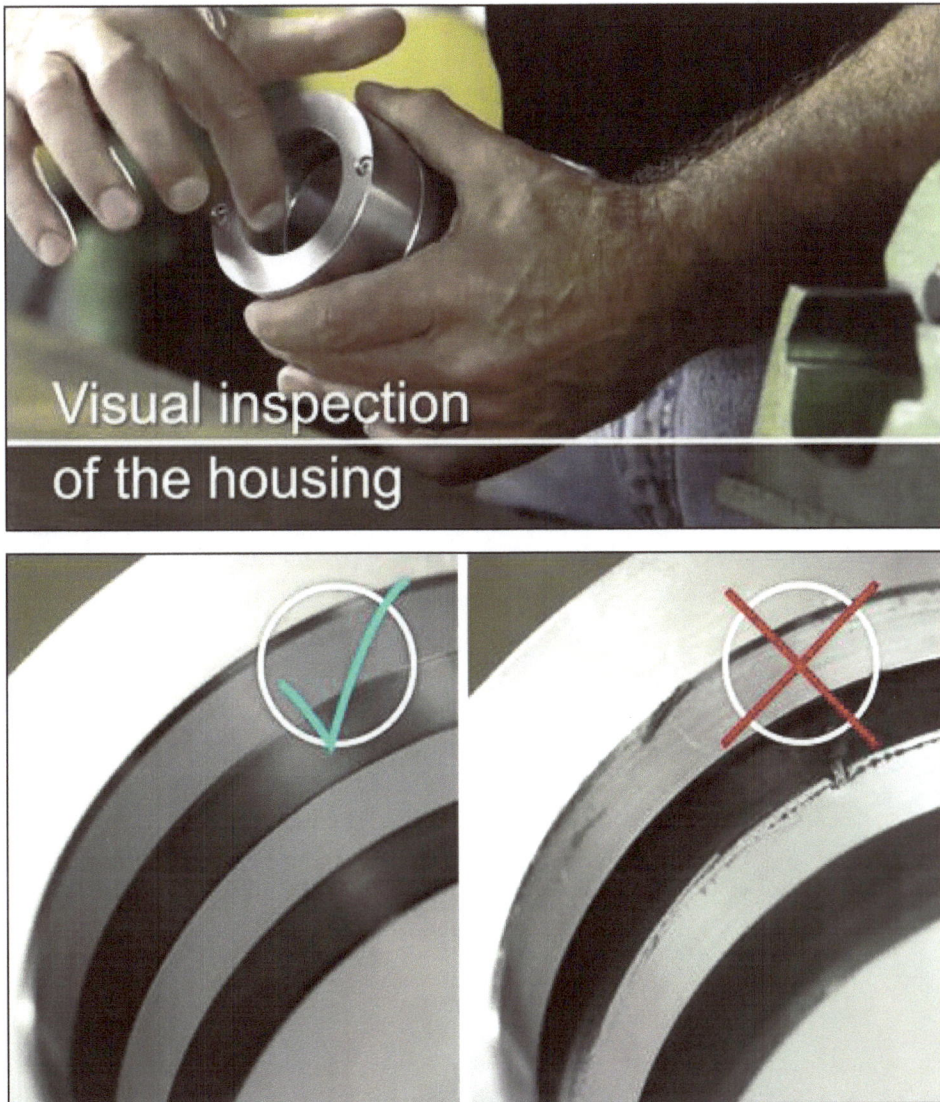

Fig. A.125 - Inspect Seal Groove (Courtesy of Trelleborg)

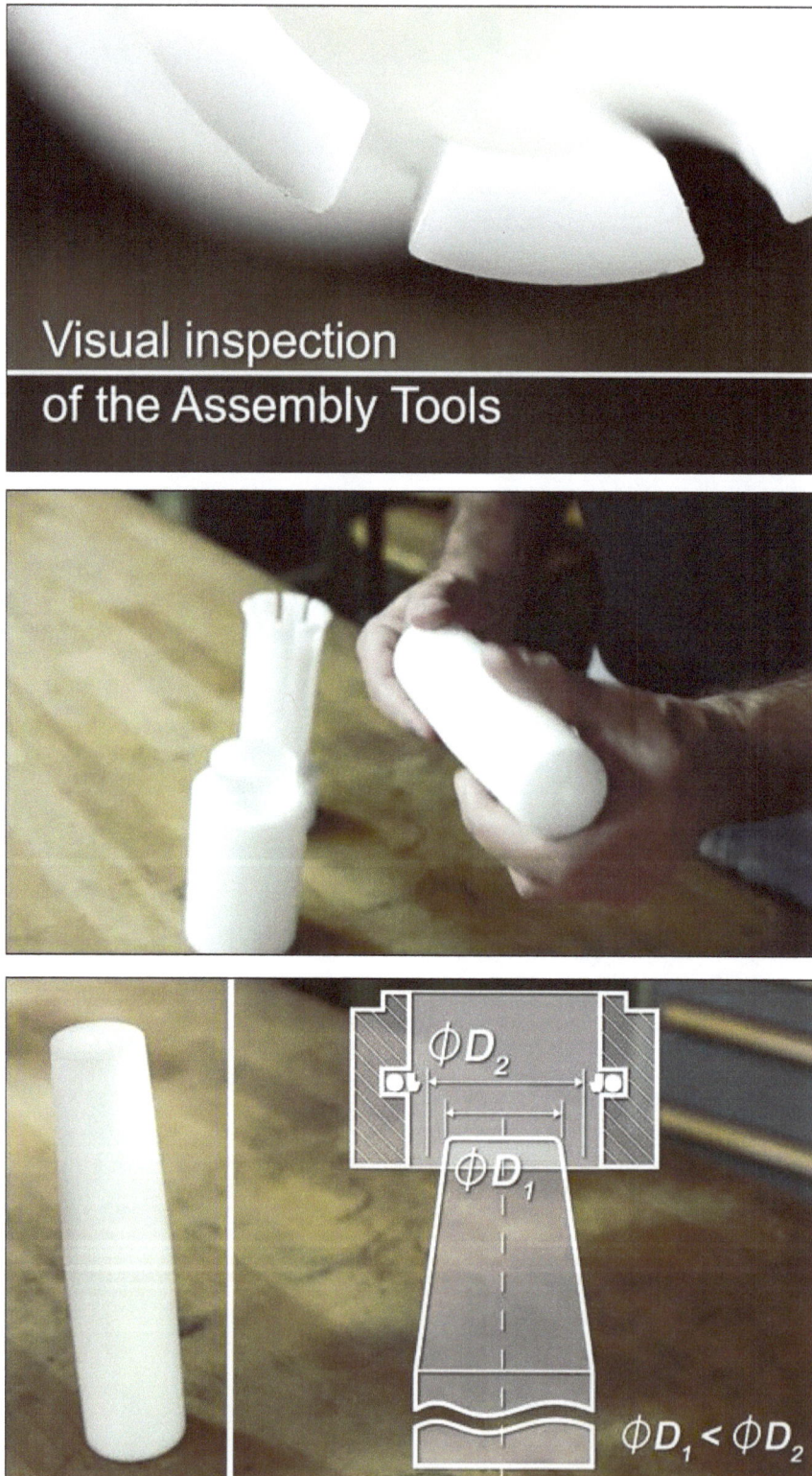

Fig. A.126 - Inspect Assembly Tools (Courtesy of Trelleborg)

Fig. A.127 - Cover Threads (Courtesy of Trelleborg)

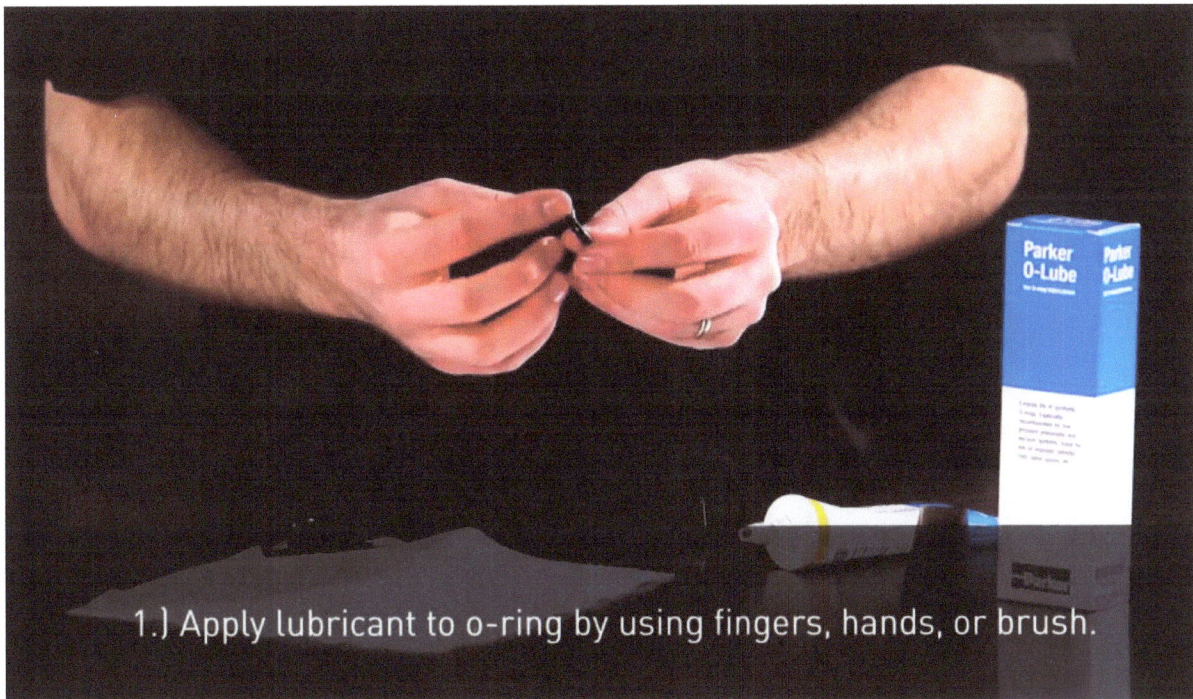

1.) Apply lubricant to o-ring by using fingers, hands, or brush.

Fig. A.128 - Lubricate Seal Before Assembly (Courtesy of Parker)

Fig. A.129 - Use Adequate Installation Tools for Rod Seals (Courtesy of Trelleborg)

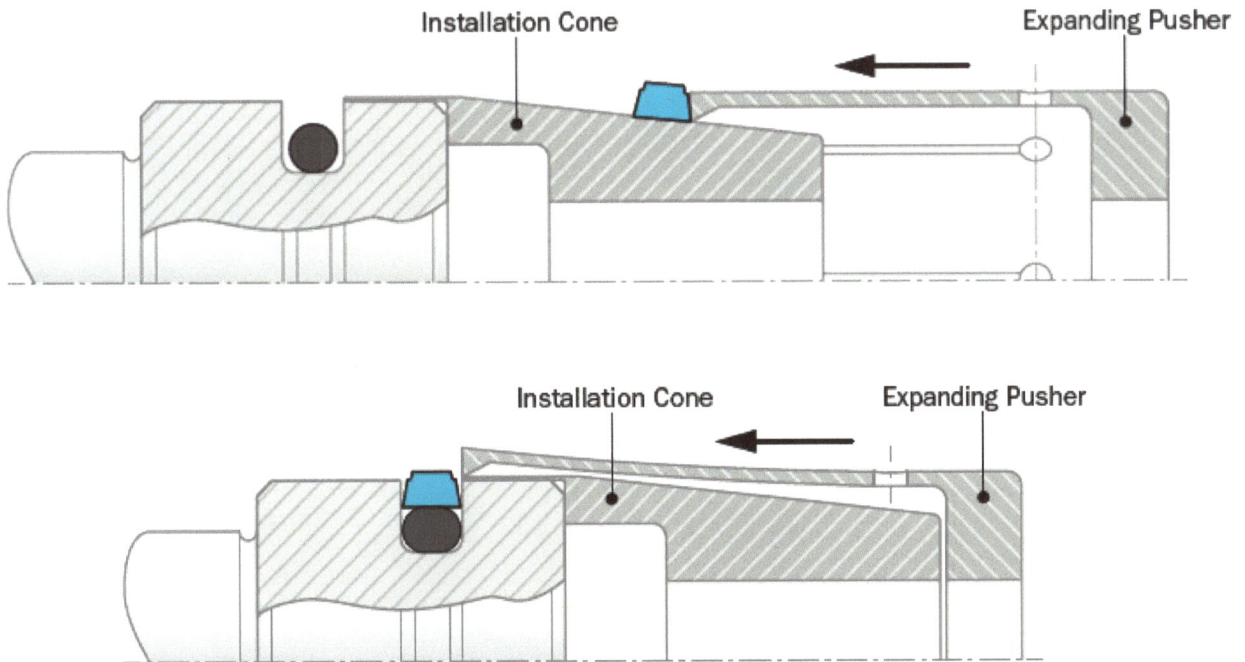

Fig. A.130 - Use Adequate Installation Tools for Piston Seals (Courtesy of Trelleborg)

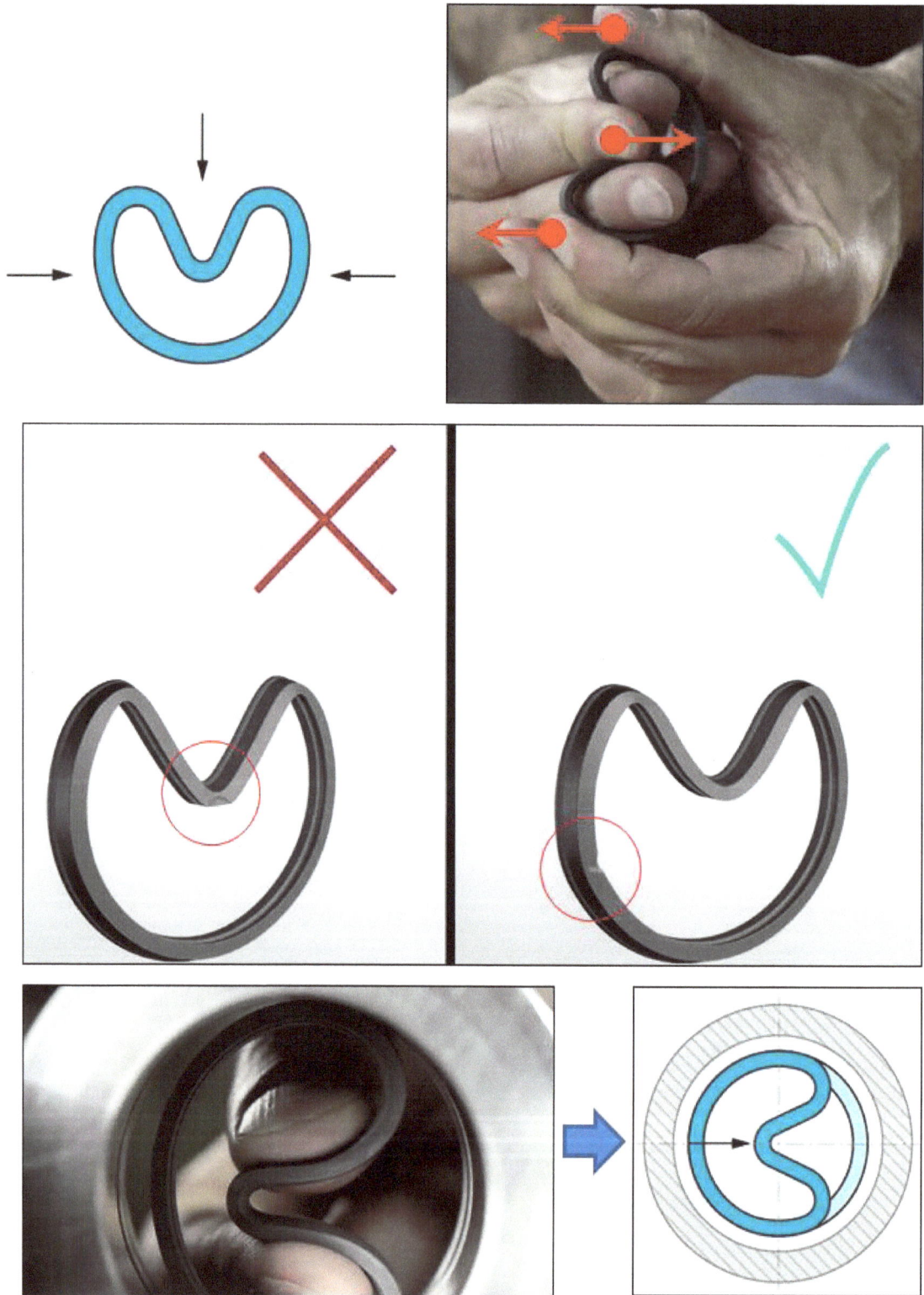

Fig. A.131 - Compress the Seal in kidney Shape (Courtesy of Trelleborg)

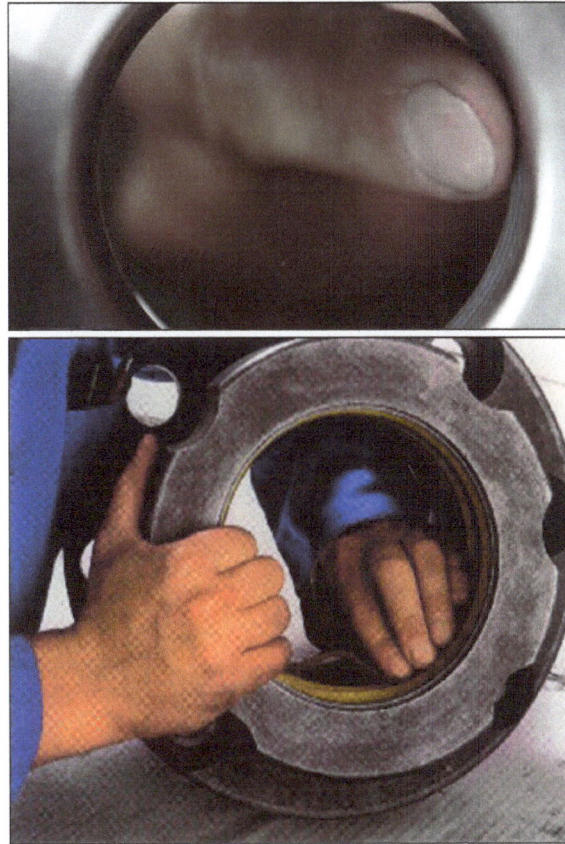

Fig. A.132 - Check Proper Installation of the Seal

Fig. A.133 - Squeeze the Seal in its Groove (Courtesy of Trelleborg)

Chapter 10 – Best Practices for Hydraulic Seals Storage

Fundamental instructions on storage, cleaning and maintenance of elastomeric seal elements is described in international standards, such as: **DIN 7716 / BS 3F68: 1977, ISO 2230, or DIN 9088**. Properties of hydraulic sealing materials are affected by the storage environment. The following list provides guidelines for storing hydraulic seals.

Best Practices for Hydraulic Seals Storage:
1. **General Conditions:** Storage space should be kept cool, dry, dust free, and moderately ventilated.
2. **Humidity:** Optimum humidity is 40 to 65 percent, maximum 75%.
3. **Temperature:** Optimum temperature is 25 OC (77 OF), maximum 50 OC (122 OF). When taken from low temperatures, items should be raised to approximately 30ºC (86ºF) before they are used.
4. **Air:** Avoid exposure to direct and continuous stream of conditioned air.
5. **Heat:** Avoid exposure to direct heat source such as boilers or radiators.
6. **Light:** Avoid exposure to direct sunlight and ultraviolet light. Unless packed in opaque containers, it is advisable to cover windows with red or orange screens or coatings.
7. **Radiation:** Avoid exposure to Gamma radiation, otherwise seals compression set is severely affected.
8. **Ozone and Oxygen:** Avoid exposure to sources of ozone such as mercury vapor lamps, high-voltage electrical equipment, combustible gases, and organic vapors. Wrapping in airtight containers, or other suitable means should be used for vulcanized rubber items. Storage in containers that limit exposure to environmental conditions (e.g. sealed plastic bags) should be used for all materials when possible.
9. **Liquids:** Avoid exposure to vapors of gasoline, greases, acids, cleaning liquids (unless such liquids are part of the seals' design or manufacturer's packaging). No solvents, fuels, lubricants or cleaning agents, or similar products, are to be stored in the same area.
10. **Contact with Elastomers:** Avoid contact between seals made from dissimilar compounds. in such cases, seals should be individually packaged.
11. **Contact with Metals:** Avoid contact with certain metals that have degrading effects on some elastomers such as manganese, iron and particularly copper.
12. **Packaging:** Pack the seals in stress-free cases and DO NOT squeeze a hydraulic seal to accommodate it in a small storage area.
13. **Deformation:** DO NOT store seals on top of each other or place heavy objects on top of any stored seals. Where possible, rubber items should be stored in a relaxed position, free from tension or compression. Laying the item flat avoiding crushing keeps it free from strain and minimizes deformation.
14. **Hanging:** DO NOT hang or suspend seals on a hook in a vertical position as gravity will distort the seal over time.
15. **Cleaning:** Organic solvents such as trichloroethylene, carbon tetrachloride, and petroleum are the most harmful agents. Soap, water, and methylated spirits are the least harmful, and all parts should be dried at room temperature before use.

16. **Stock rotation (FIFO):** Stock the seals in rotation, i.e. First-In, First-Out manner (FIFO). This ensures that the next seal used in the rotation will be within its intended shelf life.

17. **Shelf Life:** Considerable storage life without detectable damage varies for different elastomers. In 1998, the Society of *Automotive Engineers* (SAE) issued an *Aerospace Recommended Practice* (ARP) for the storage of elastomer seals and seal assemblies prior to installation. Table A.12 shows approximate shelf life for standard elastomers.

Compound Name	ASTM Polymer	Shelf Life
Aflas®	FEPM	Unlimited
Butyl Rubber, Isobutylene Isoprene	IIR	Unlimited
Chloroprene (Neoprene®)	CR	15 Years
Chlorosulphonated Polyethylene (Hypalon®)	CSM	15 Years
Epichlorohydrin (Hydrin®)	ECO	NA
Ethylene Acrylic (Vamac®)	AEM	15 Years
Ethelene Propylene, EPDM or LP	EP	Unlimited
Fluorocarbon (Viton®)	FKM	Unlimited
Fluorosilicone	FVMQ	Unlimited
Hydrogenated Nitrile, HNBR or HSN	HNBR	15 Years
Nitrile (BUNA-N or NBR)	NBR	15 Years
Perfluoroelastomer	FFKM	Unlimited
Polyacrylate	ACM	15 Years
Polyurethane (Polyester or Polyether)	AU/EU	5 Years
Silicone	Q,VMQ,PVMQ	Unlimited
Styrene Butadiene (Buna-S)	SBR	3 Years

Table A.12 - Approximate Shelf Life for Standard Elastomers (Courtesy of MFP Seals)

Chapter 11 – Best Practices for Hydraulic Seals Failure Analysis

Challenges of Seals Failure:

- **Possible Remedies:** Seal damage is irreversible and there are no remedies for failed seals.

- **Consequences of Seal Failure:** The consequences of a seal failure may vary from a simple internal or external leakage to a bigger problem. A critical seal would be one that if failure occurs would create a hazard for personnel and/or the public. Application examples where critical seals are used are aircraft, amusement park rides, and elevating devices.

- **Cost of the seal vs. the Consequences:** The cost of hydraulic seals is pennies compared to the time and resources put into disassembling components to merely inspect the seals.

- **Failure Analysis:** Without laboratory analysis, it may not be easy by just visual inspection to determine the condition of seal deterioration over the time.

As shown in Fig. A.134, failure modes of hydraulic sealing elements can be categorized as follows:

A. Manufacturing Defects.
B. Seal and Gland (Groove) Design Defects.
C. Assembly Defects.
D. Operational Defects.
E. Normal Aging Defects.
F. Storage Defects.

To make it easy to jump to a specific failure mode, the following subtitles are numbered based on the number of the failure mode on the chart.

Hydraulic Seals Failure Analysis

A- Manufacturing Defects

1-Improper Molding

2-Poor Material

B- Design Defects

3-Extrusion

4-Gland Sharp Corner

5-Rough Interface Surfaces

6-Blow-By Effect

C- Assembly Defect

7-Passing Over Sharp Edges

D- Operational Defect

8-Fluid Overpressure

9-Pressure Trapping

10-Fluid Overheating

11-Overspeeding

12-Fluid Contamination

13-Fluid Incompatibility

14-Fluid Chemical Attack

15-Hydrolysis

16-Explosive Decompression

17-Dieseling

18-Side Loading

19-Vibration

20-Spiral Failure

21-Seal Wear

22-Fatigue

E- Normal Aging Defects

23-Hardening

24-Splits

F- Storage Defects

25-Swelling

26-Ozone Cracking

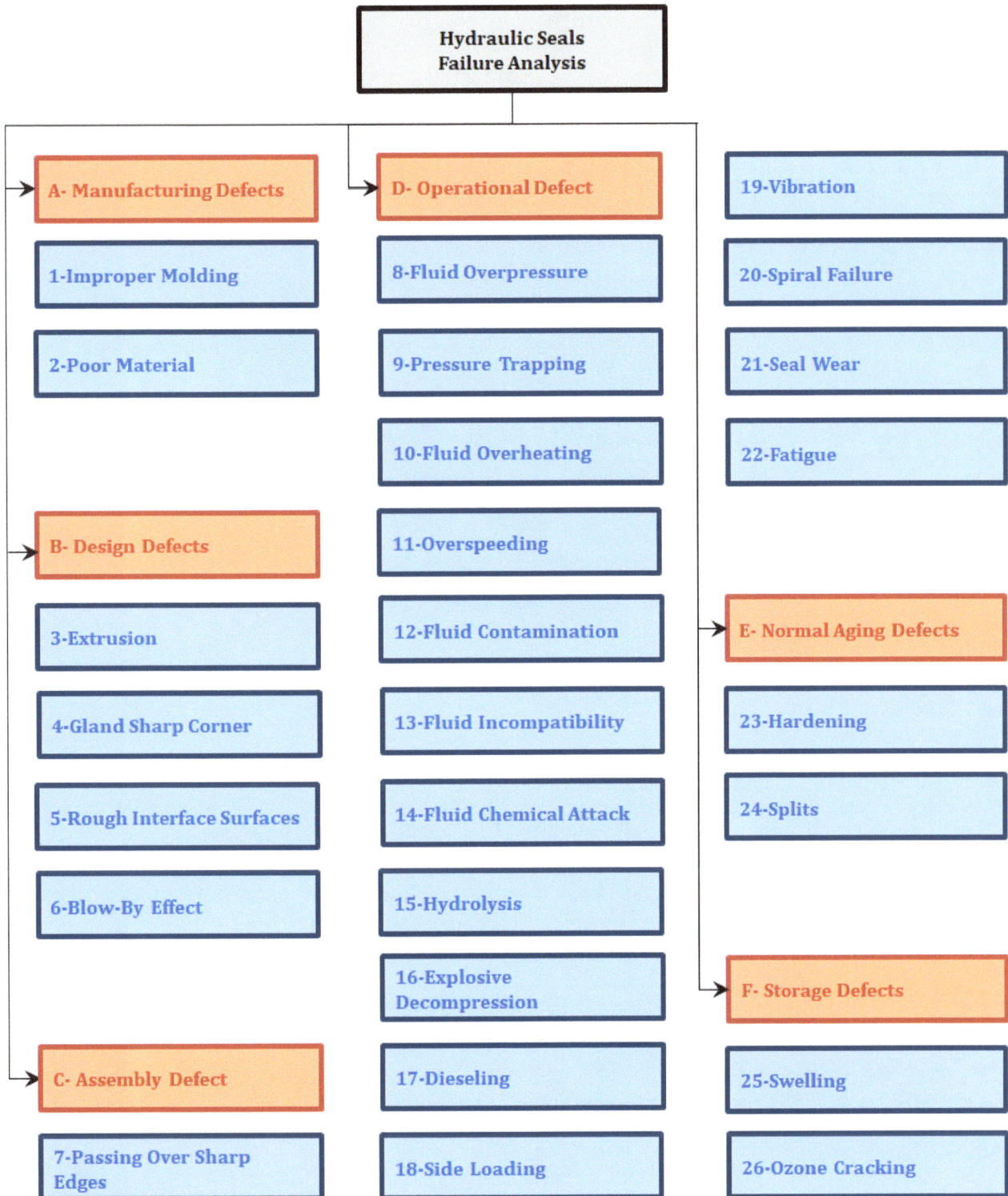

Fig. A.134 - Hydraulic Seal Failure Analysis Diagram

11.1- Manufacturing Defects - Improper Molding

Failure Source: Improper *Molding* process due to defected dies, improper injection flow, pressure or temperature.

Failure Mode: As shown in Fig. A.135, material of a hydraulic seal is harshly scorched.

Suggested Solution: Review molding process and conditions.

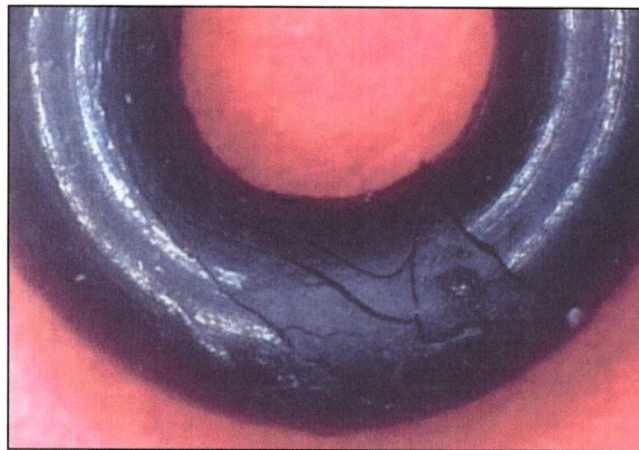

Fig. A.135 - O-Ring Failure due to Improper Molding (www.o-ring-lab.com)

11.2- Manufacturing Defects - Insufficient Material Properties

Failure Source: Severe compression set due to poor material properties.

Failure Mode: As shown in Fig. A.136, a hydraulic seal is permanently deformed in short time.

Suggested Solution: Select better quality hydraulic seal.

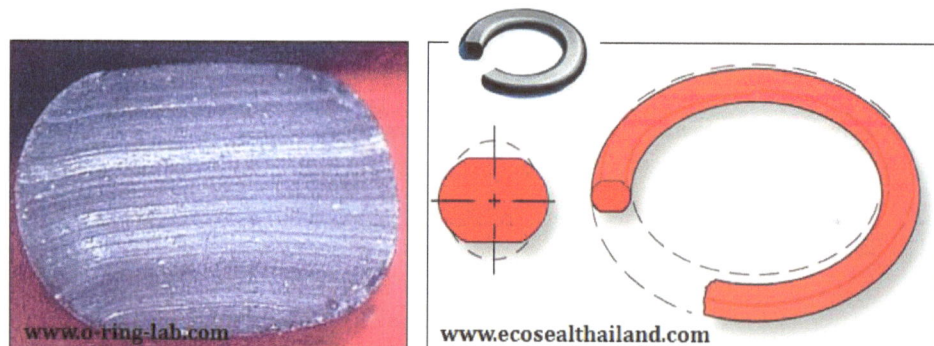

Fig. A.136 - O-Ring Failure due to Insufficient Material Properties

11.3- Design Defects - Extrusion

Failure Source: extrusion and nibbling are caused by one or more of the following conditions:
- **Excessive Sealing Gap:** Excessive clearances (sealing gap) Or irregular clearance gaps caused by eccentricity.
- **Overpressure:** Working pressure exceeds the recommended limits.
- **Less Hardness:** Seal material is too soft.
- **Degradation:** swelling, softening, shrinking, cracking, etc.)
- **Improper Size:** Too large seal installed causing excessive filling of groove.

Extrusion Limits are defined based on the seal hardness "Shore A", diametral clearance, and the working pressure. Figure A.137 shows the extrusion Limits. As shown in the figure, at the same working pressure (red arrows), the sealing gap (diametral clearance) and extrusion limits are directly proportional to the seal hardness. On the other hand, for the same seal hardness (blue arrows), the sealing gap and extrusion limit are inversely proportional to the working pressure.

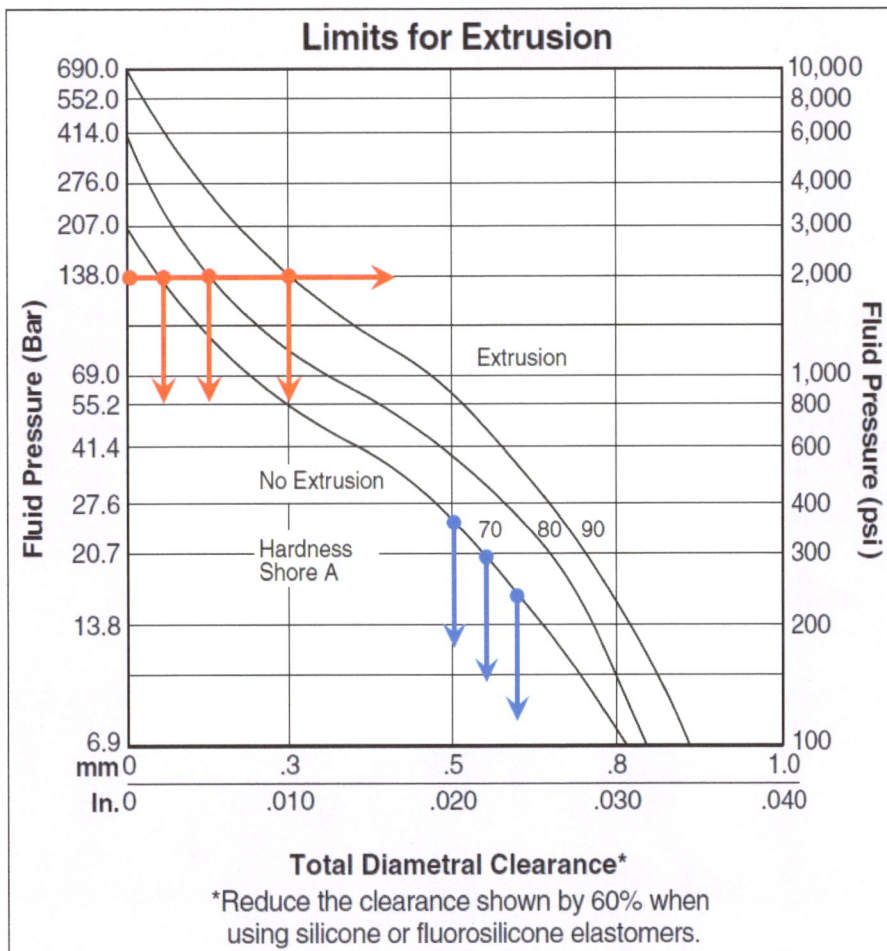

Fig. A.137 - Limits of Extrusion (Courtesy of Parker)

Failure Mode:

If Extrusion Limits are exceeded, as shown in Fig. A.138, the O-Ring will extrude into the sealing gap.

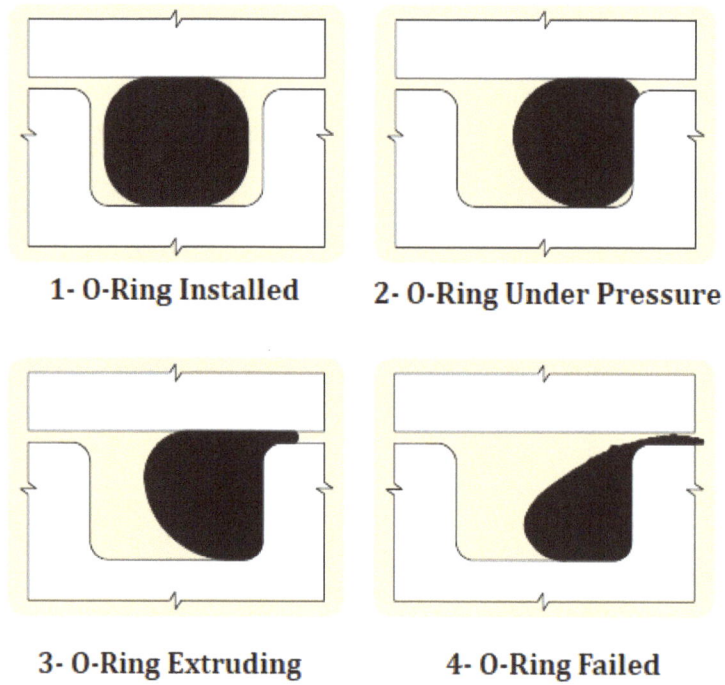

1- O-Ring Installed   2- O-Ring Under Pressure

3- O-Ring Extruding   4- O-Ring Failed

Fig. A.138 - O-Ring Extrusion due to Excessive Seal Gab (Courtesy of Parker)

Figures A.139 through A.141 show hydraulic seals creeping into the sealing gap. As a result, the seal deforms and/or breaks off.

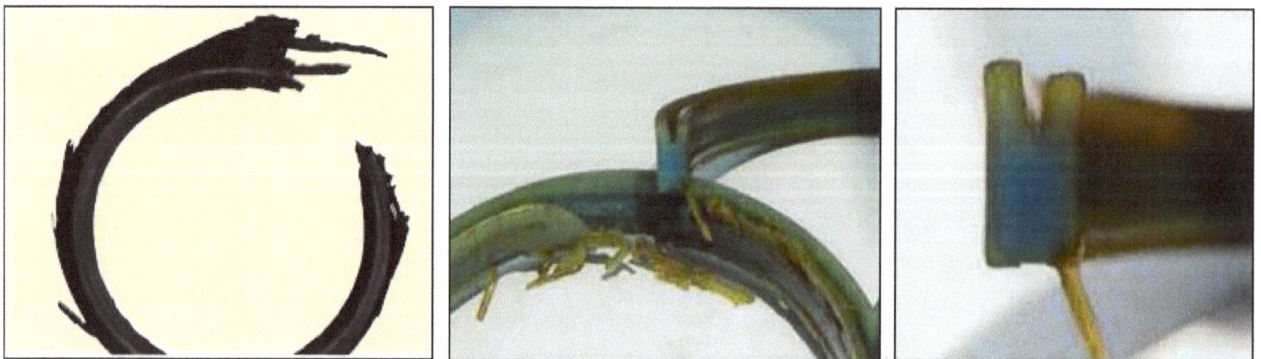

Fig. A.139 - Seal Extrusion, Example 1 (Courtesy of Parker)

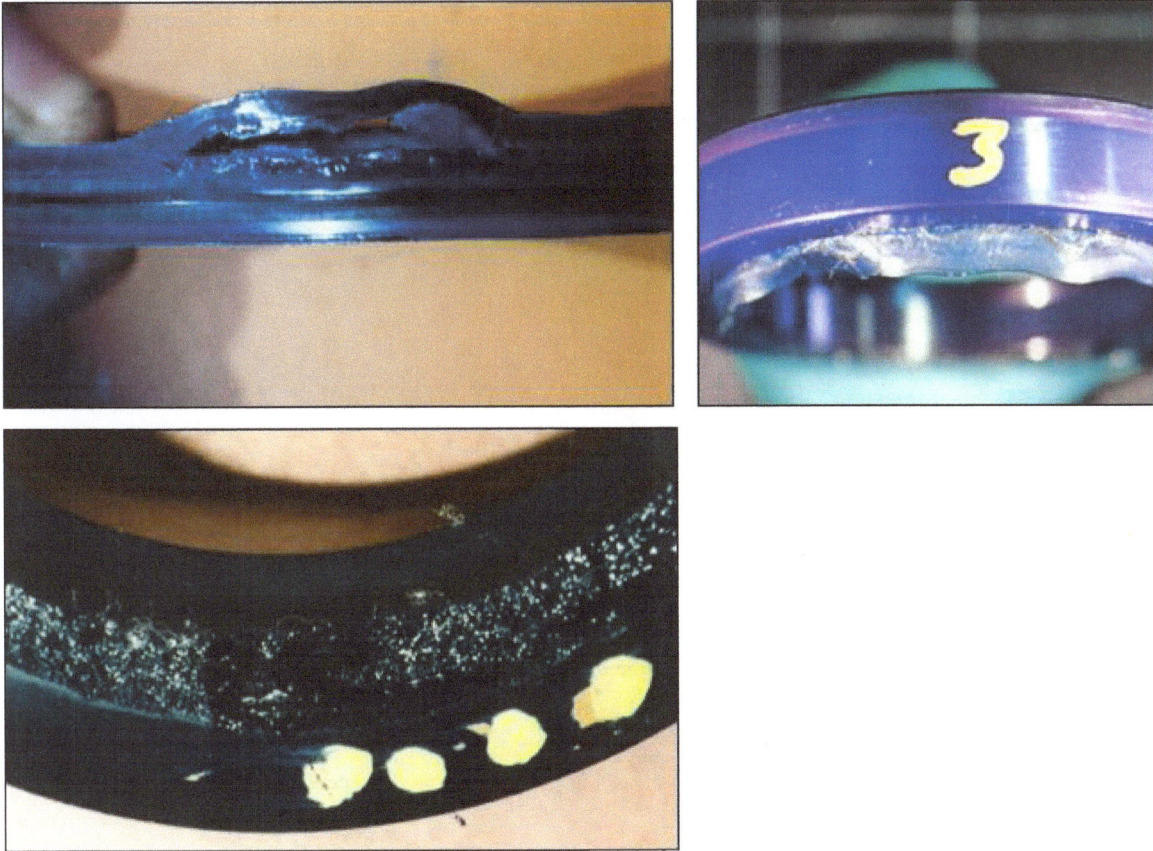

Fig. A.140 - Seal Extrusion, Example 2

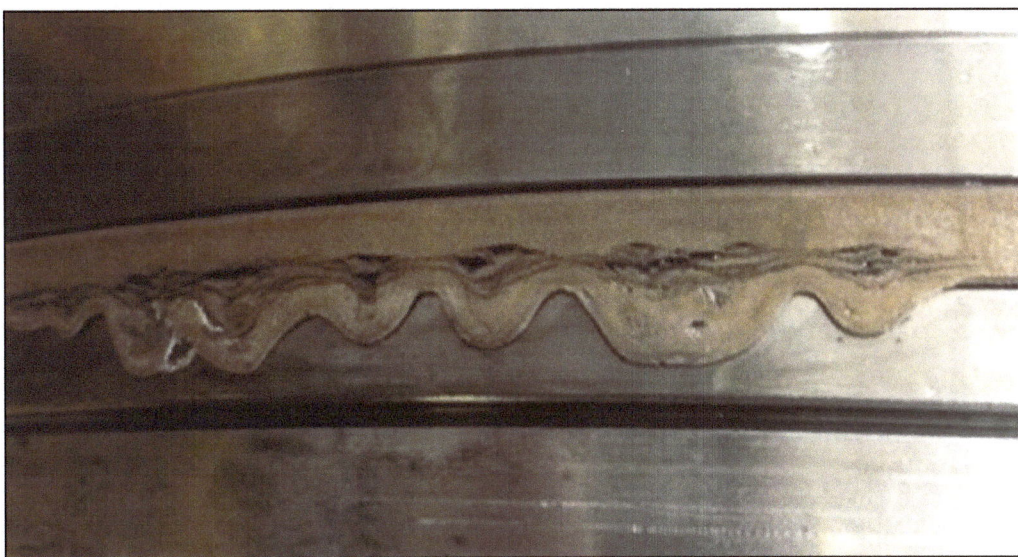

PTFE piston seal showing signs of severe extrusion

Fig. A.141 - Seal Extrusion, Example 3 (Courtesy of System Seals Inc.)

Standard Test Method - Extrusion-Resistance Test (ASTM C1183 / C1183M): this method is used to determine the ability of a hydraulic seal to resist extrusion under certain pressure.

Suggested Solution: Seal gap extrusion can be avoided by considering the following solutions:

Solution 1-Proper Design of Sealing Gap:

A sealing gap must be properly designed to avoid seal extrusion.

Solution 2-Use of Back-up Rings:
As shown in Fig. A.142, backup rings are placed on the back side of O-Rings or dynamic seals to support the seal against extrusion.

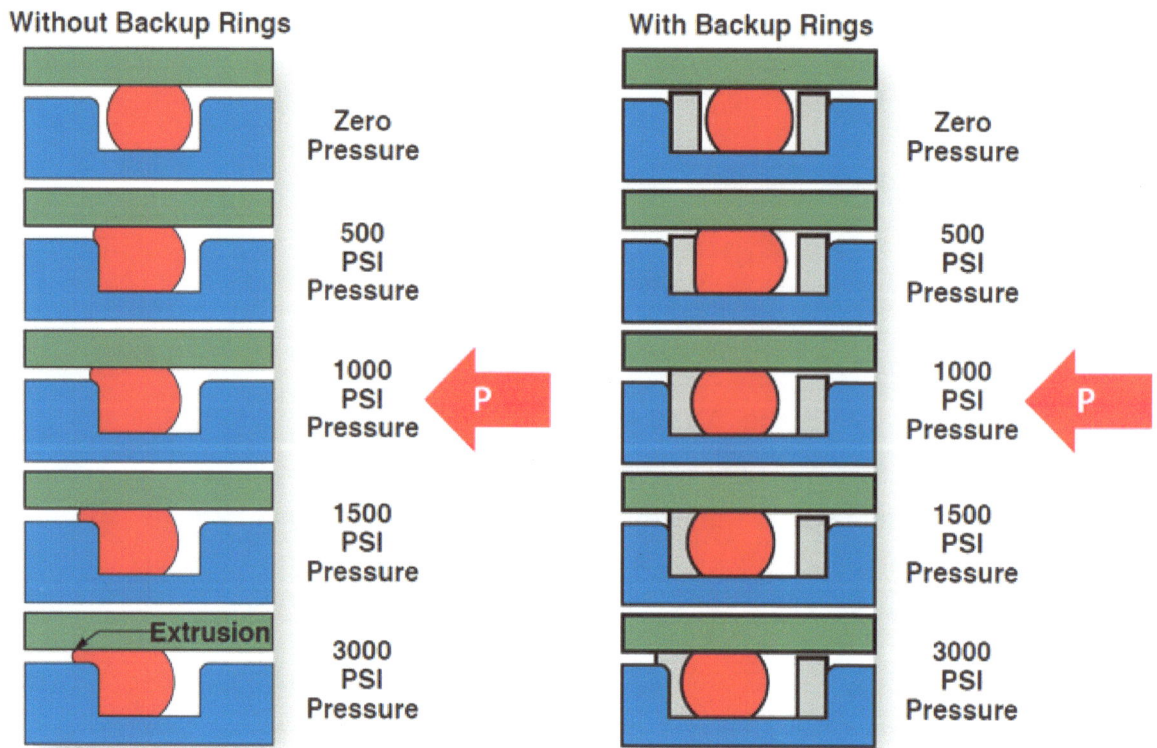

Fig. A.142 - Seal Extrusion Avoidance by use of Back-up Rings (www.ecosealthailand.com)

Solution 3-Use of Anti-Extrusion Wedge-Rings:

As shown in Fig. A.143, a *Wedge-Ring* is placed on the back side of an O-Ring. The Wedge-Ring adjusts the gap dynamically as the pressure changes.

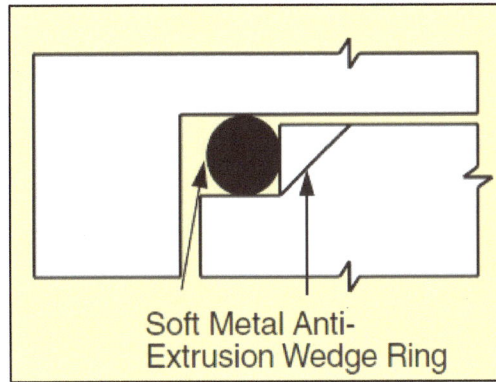

Soft Metal Anti-Extrusion Wedge Ring

Fig. A.143 - Seal Extrusion Avoidance by use of Wedge-Rings (Courtesy of Parker)

11.4- Design Defects - Gland (Groove) Sharp Corners

Failure Source: Improper groove design with sharp corners.

Failure Mode: A seal groove is not adequately rounded. As shown in Fig. A.144, an O-Ring was used in a stack valve at 250 bar (3,625 psi) and was damaged at the outer circumference even with relatively small seal gaps.

Suggested Solution: Design seal gland (groove) to meet the design codes.

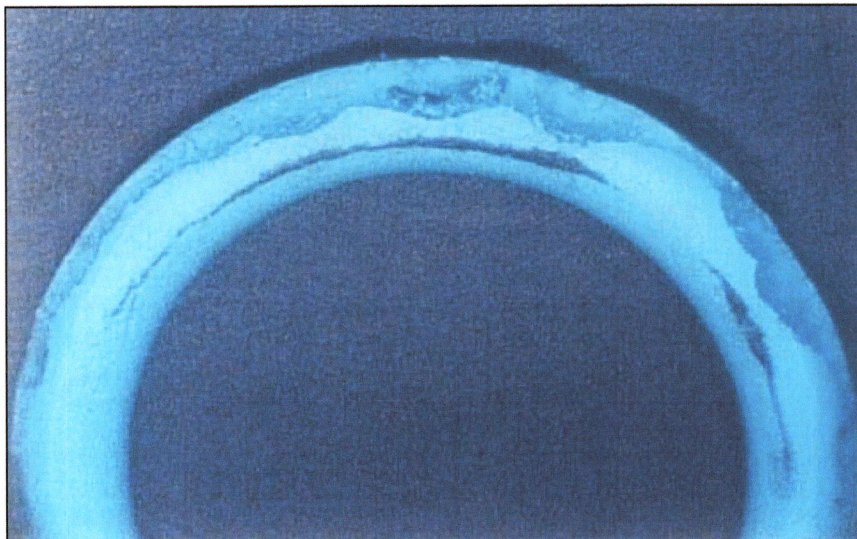

Fig. A.144 - O-Ring Failure due to Sharp Corners and High Pressure (www.o-ring-lab.com)

11.5- Design Defects - Rough Surfaces

Failure Source: Finishes of contact surfaces have much to do with the life of dynamic seals. A seal can be exposed to abrasion due to:
- Contact with very rough surface results in high friction with the seal material.
- Contact with very smooth surface results in running the seal dry without lubrication.

Failure Mode: Figures A.145 through A.148 show hydraulic seals abrasion marks.

Suggested Solution: The most desirable surface finish value is from 10 to 20 micro-inches. However, the surface must be rough enough to hold small amounts of oil. surface finish values less than 5 micro-inches are not recommended for dynamic seals. Otherwise, an extending rod will be wiped completely dry and will not be lubricated when it retracts.

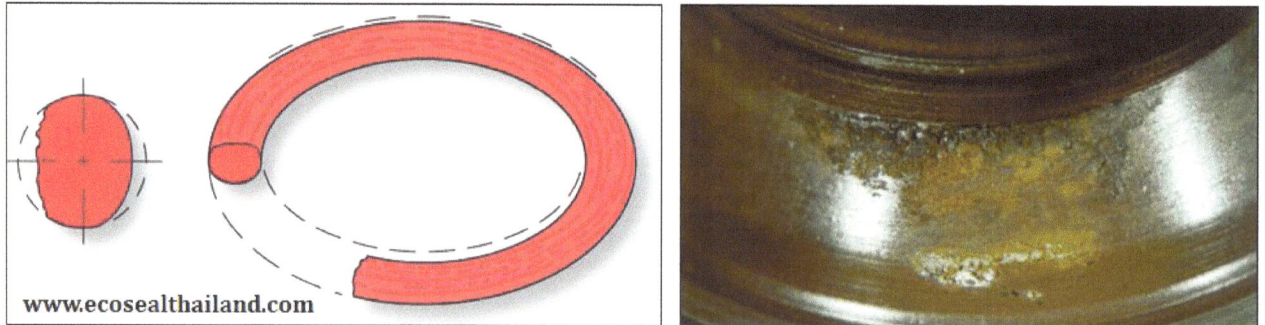

www.ecosealthailand.com

Fig. A.145 - Seal Abrasion due to Contact with Rough Surface, Example 1

NBR piston u-cup with abrasion marks at the seal lip and migrating across the dynamic surface

**Fig. A.146 - Seal Abrasion due to Contact with Rough Surface, Example 2
(Courtesy of System Seals Inc.)**

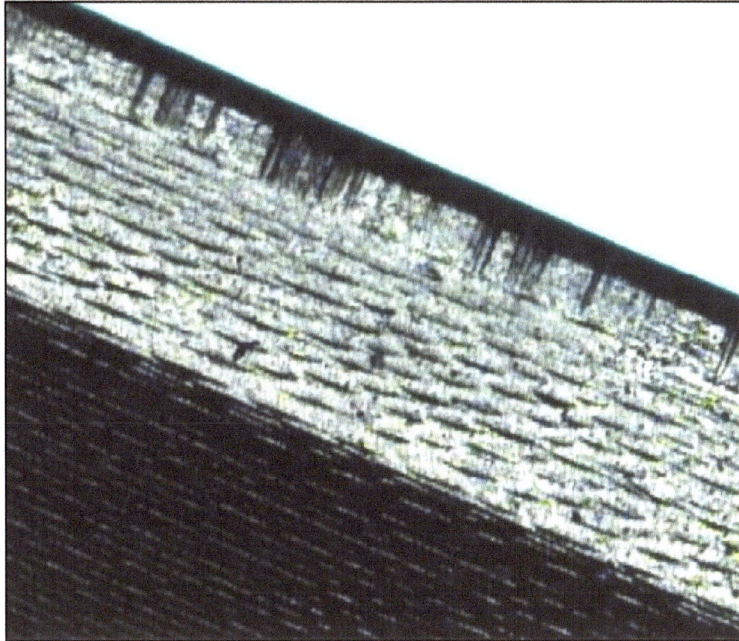

U-cup seal lip showing abrasion marks

**Fig. A.147 - Seal Abrasion due to Contact with Rough Surface, Example 3
(Courtesy of System Seals Inc.)**

PTFE seal with abrasion marks at the seal lip and migrating across the dynamic surface.

**Fig. A.148 - Seal Abrasion due to Contact with Rough Surface, Example 4
(Courtesy of System Seals Inc.)**

11.6- Design Defects - Blow-By Effect

Failure Source: Figure A.149 shows a traditional design of a bidirectional piston sealing solution. In such design, *Blow-By Effect* occurs causing leakage.

Failure Mode: Increased leakage rate.

Suggested Solution: As shown in Fig. A.150, to resolve this problem, piston seal should contain radial grooves or notches. One notch for a unidirectional seal or two notches for a bidirectional seal. With such notches, fluid pressure reaches out to the O-Ring and compresses it in a way that controls dynamically the clearance between the piston seal and the cylinder wall. As a result, leakage across the piston is controlled.

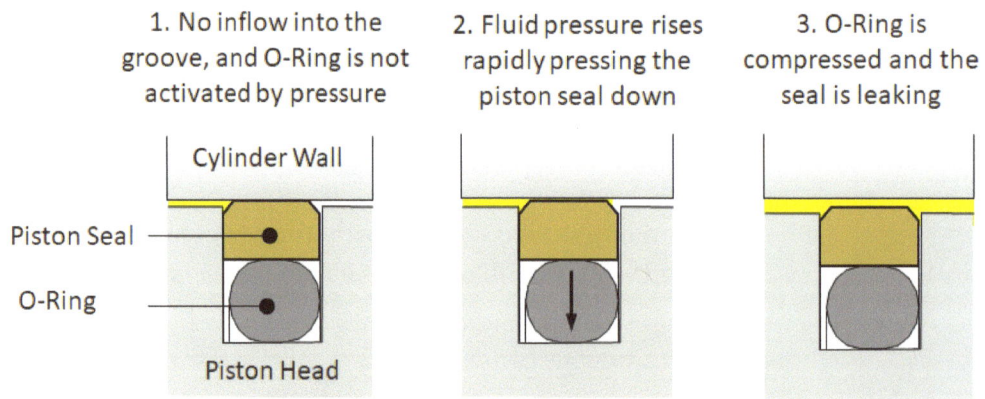

Fig. A.149 - Blow-By Effect

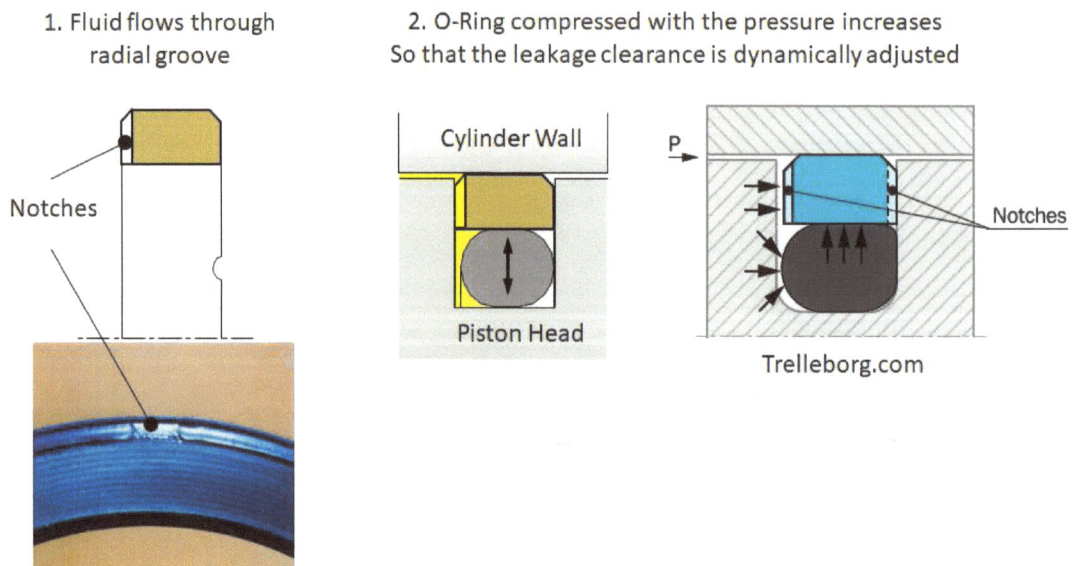

Fig. A.150 - Resolving Blow-By Effect

11.7- Assembly Defects - Passing Over Sharp Edges

Failure Source: Damage to hydraulic seals can occur during installation when:
- Seals come in contact with sharp edges such as threads.
- Insufficient lead-in chamfer.
- Oversize piston seal.
- Undersize rod seal.
- Seal is twisted/pinched during installation.
- Seal is not properly lubricated before installation.
- Seal is dirty or contaminated with metal particles upon installation.

Failure Mode: As shown in Fig. A.151, failure can occur due to passing over sharp edges in the gland area. As shown in Fig. A.152, short cuts can happen due to passing over sharp edges in the lead-in chamfer or sharp threads.

Suggested Solution:
- Cover threads and sharp edges before assembly.
- Use of proper installation tools.
- Make sure lead in chamfers are based on manufacturers recommendations.
- Select proper seal size.
- Consider proper cleanliness during assembly.

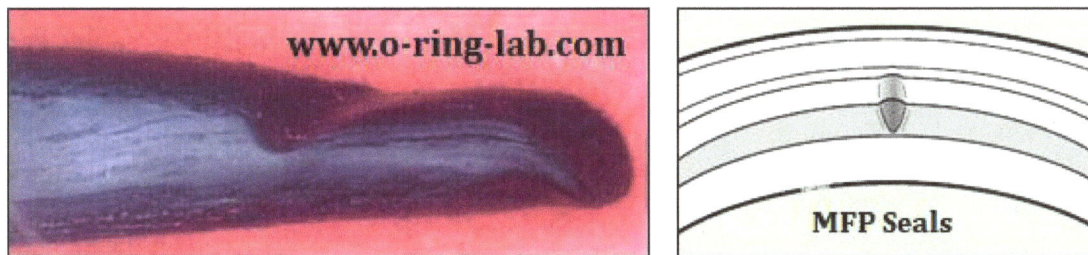

Fig. A.151 - Sheared O-Ring due to Passing Over Sharp Edges in the Gland Area During Assembly (www.o-ring-lab.com)

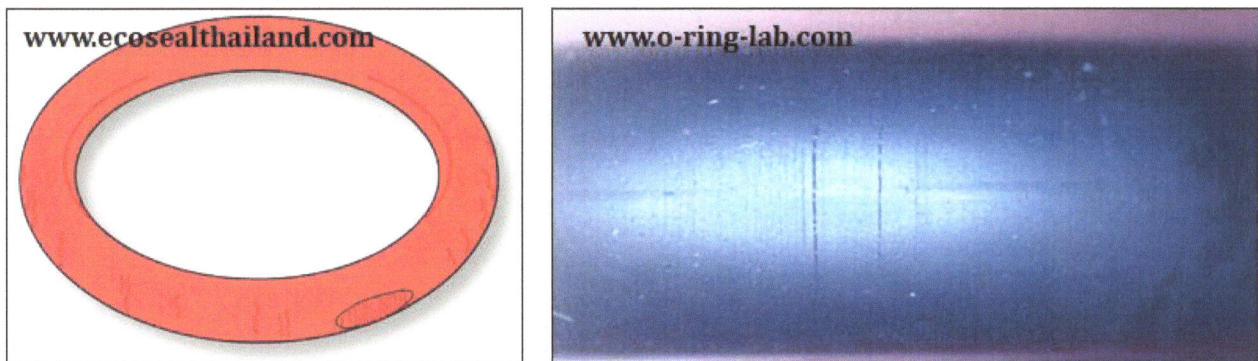

Fig. A.152 - O-Ring Failure due to Passing Over Guide Chamfer

11.8- Operational Defects - Overpressure

Failure Source: overpressure work environment.

Failure Mode: As shown in Figures A.153 and A.154, hydraulic seals are stressed beyond its limits and fails

Suggested Solution:
- Work within recommended pressure range.
- Select seal material based on maximum working pressure

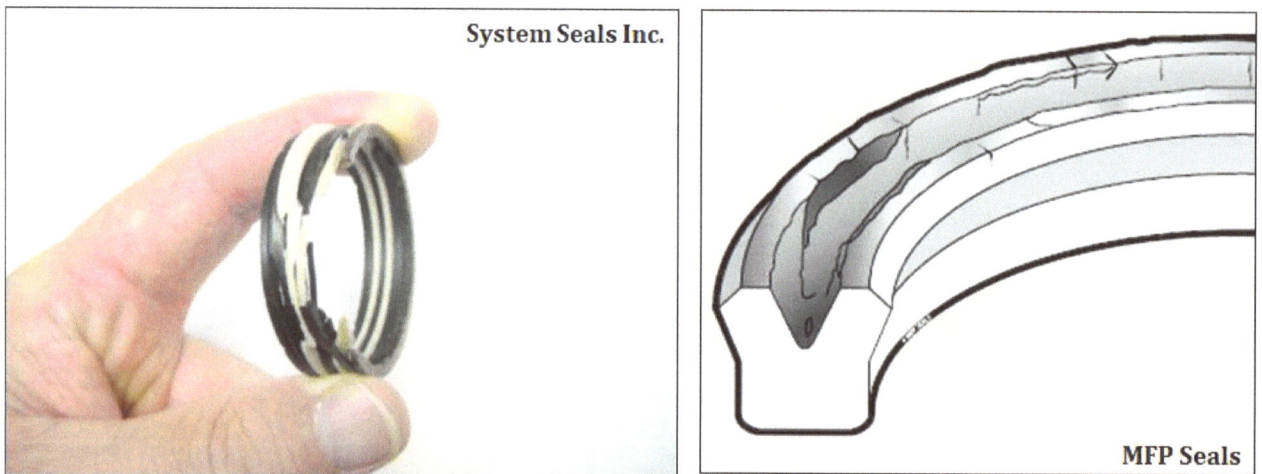

V-packing set that structurally failed at high pressure

Fig. A.153 - Seal Failure due Overpressure, Example 1

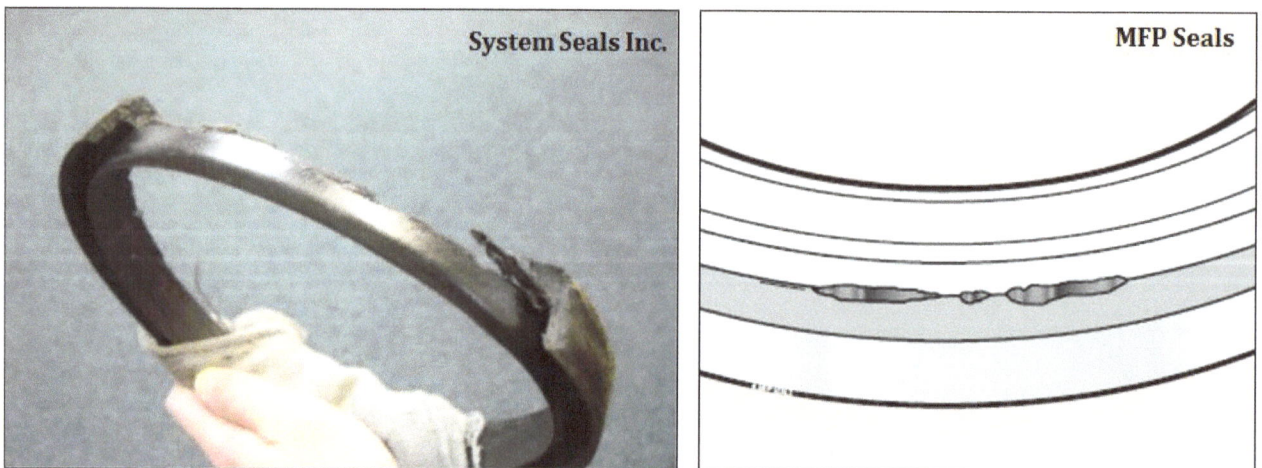

Rubber and fabric piston seal that failed from over-pressurization

Fig. A.154 - Seal Failure due Overpressure, Example 2

11.9- Operational Defects - Pressure Trapping

Failure Source: Pressure spikes, such as those created by sudden shifting of a directional valve, may be ten times greater than the normal operating pressure of a hydraulic seal. If pressure spikes occur often, as shown in Fig. A.155, pressure is trapped between seals causing seal damage.

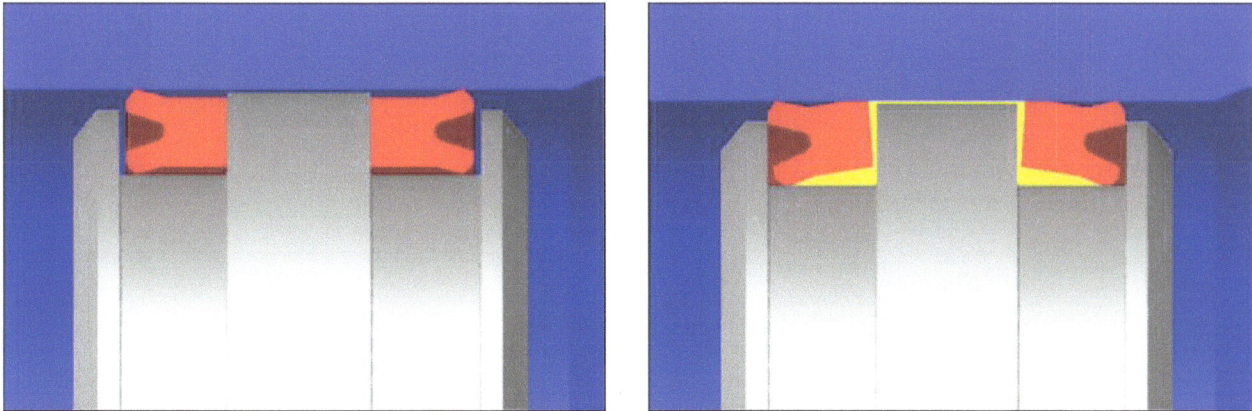

Normal action of piston can cause certain types of piston seals to trap pressure. Excessive pressure between the seals can push the seals away from each other, ultimately resulting in pressure trapping failure

Fig. A.155 - Bidirectional Seal Failure due to Overpressure (Courtesy of System Seals Inc.)

Failure Mode: As shown in Figures A.156, a hydraulic seal is squeezed and failed as a result of *Pressure Trapping*.

Suggested Solution: Apply design strategies to eliminate or at least minimize creating pressure shocks.

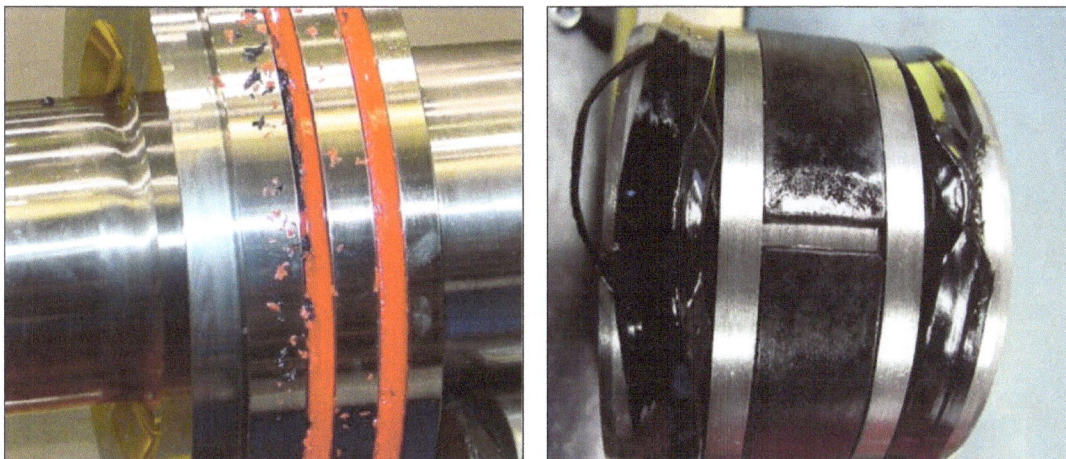

Left: Back-to-back piston u-cup seals showing pressure trapping with reverse extrusion as a result
Right: Loaded polyurethane u-cups showing severe pressure trapping and failure.

Fig. A.156 - Seal Failure due Pressure Trapping (Courtesy of System Seals Inc.)

11.10- Operational Defects - Overheating

Failure Source:
- Long exposure to high working temperature or excessive heat.
- High speed operation that increases the seal lip temperature.

Failure Mode: Heat generated by friction or directly through other heat sources accelerates hardening of the seal material, particularly in the contact area between the sealing lip and the sliding surface. This leads to cracks which become increasingly larger over time and ultimately result in seal failure. Figures A.157 through A.160 shows various failure modes such as: hardening and cracks of the sealing lip, softening permanent deformation of the seal body, extrusion, splits, ruptures, melting, and squeezing. It is to be noted also that ow temperature, beyond the seal material's specified minimum temperature, makes the seal brittle.

Suggested Solution:
- Work within recommended temperature range.
- Select proper seal material that works better at high temperature.

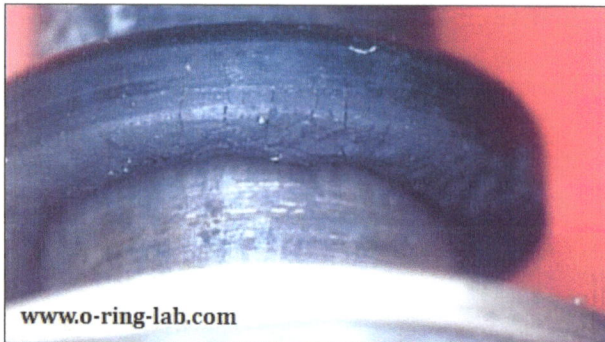

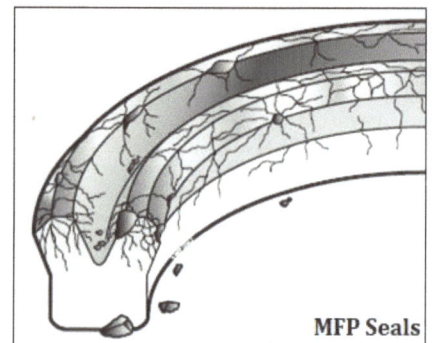

Fig. A.157 - Seals Hardened and Cracked due to Long-Term Exposure to High Temperature

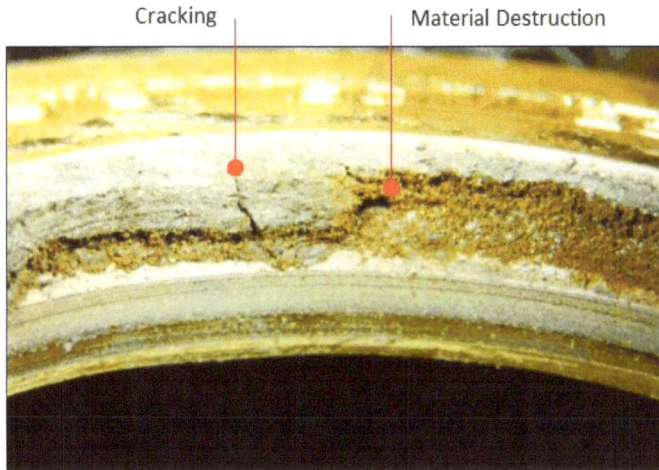

Fig. A.158 - Shaft Seal Failure due to Exposure to Excessive Working Temperature with Grease Lubricant

Fig. A.159 - V-Packing Failure due to Exposure to Excessive Working Temperature

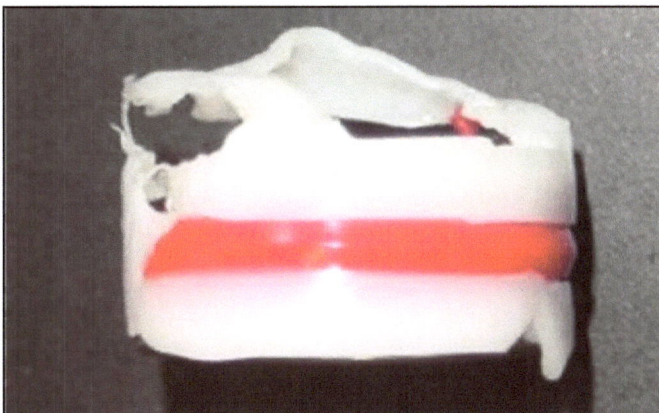

Fig. A.160 - Multi-Component Piston Seal Melted due to Exposure to Excessive Working Temperature (Courtesy of System Seals Inc.)

11.11- Operational Defects – Over speeding

Failure Source: High surface speed generates excessive heat.

Failure Mode: As shown in Fig A.161, the dynamic seal is hardened showing cracks and glazing of the seal material.

Suggested Solution:
- Reduce surface speed (stroke speed or RPM).
- Select proper seal material that works better at high speed and high temperature.

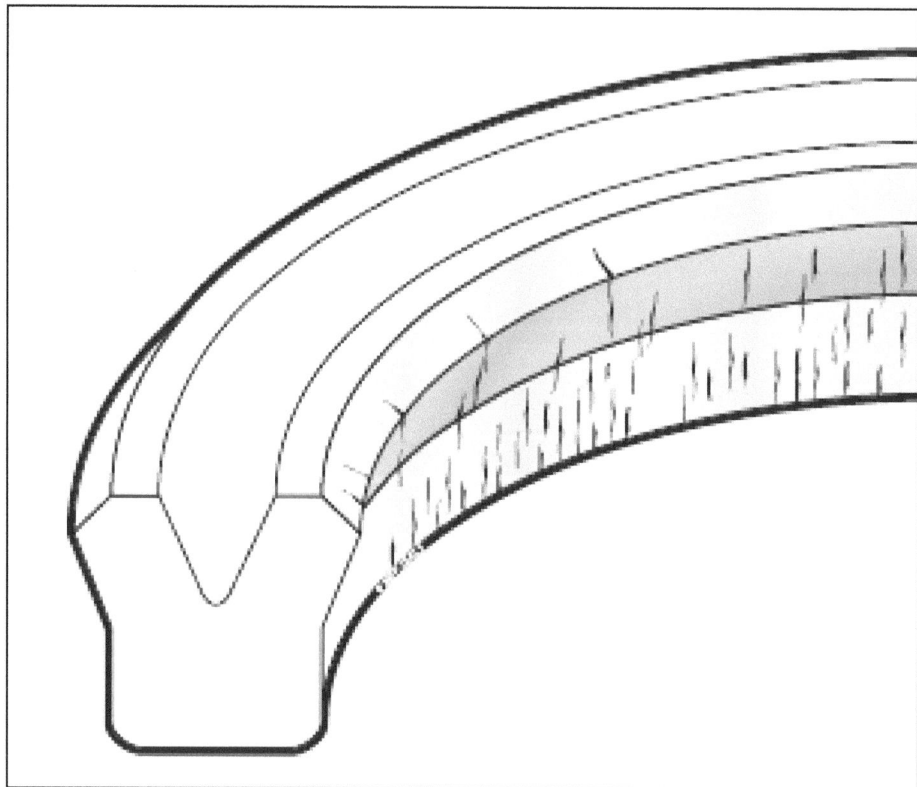

**Fig. A.161 - Hardened and Cracked Dynamic Sealing Surface due to Overspeeding
(Courtesy of MFP Seals)**

11.12- Operational Defects - Contamination

Failure Source:
- Hydraulic fluids contaminated by abrasive and metal particles.
- Dirty assembly area.
- Poor wiper performance.
- Highly contaminated environment around a cylinder as shown in Fig. A.162.

Failure Mode: Figures A.163 and A.164 show damaged hydraulic seals due to contamination.

Suggested Solution: Make sure hydraulic fluids comply with the cleanliness level recommended by the system manufacturer.

**Fig. A.162 - Hydraulic Cylinder Operating in a Severe Salt Contaminated Environment
(Courtesy of System Seals Inc.)**

The metal particles embedded in the seal produce scores on the mating surface.

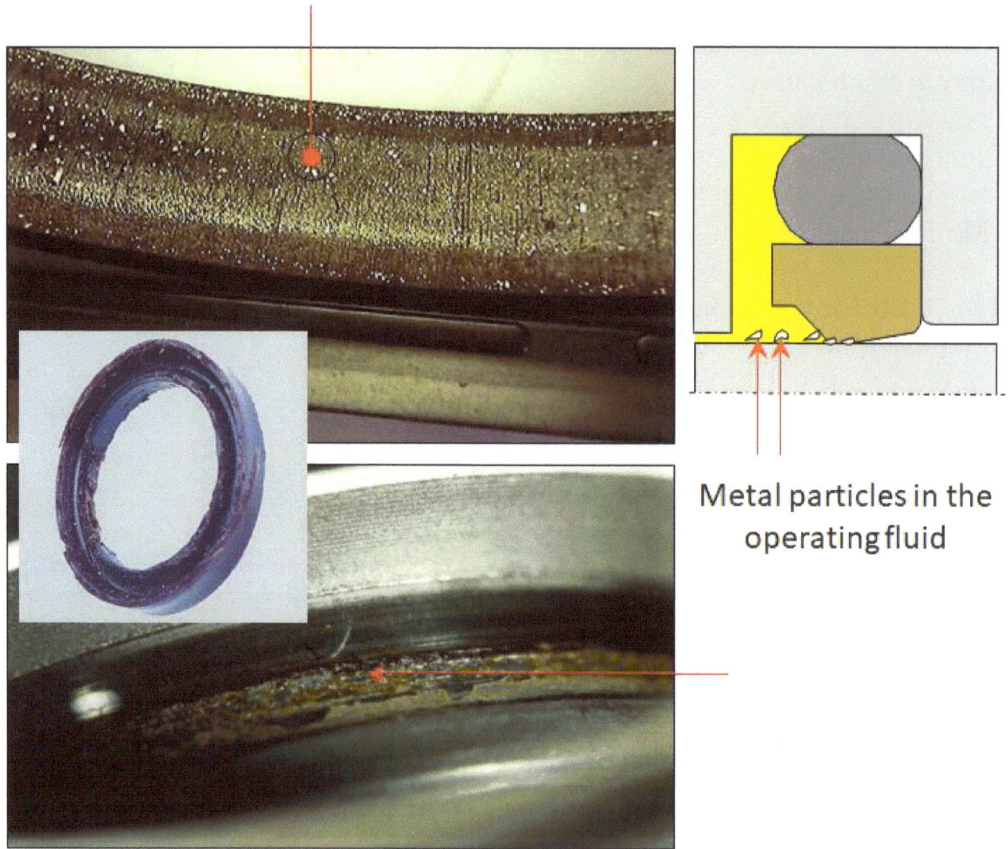

Metal particles in the operating fluid

Fig. A.163 - Seal Failure due to Contamination

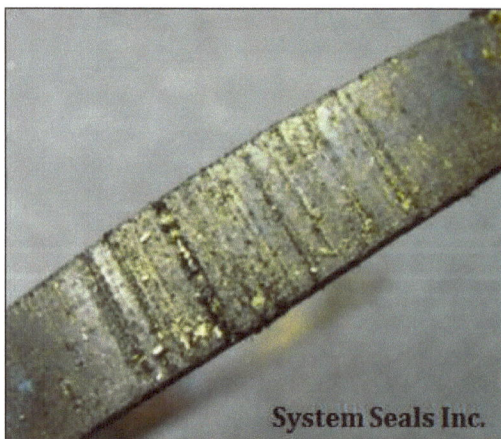

System Seals Inc.

MFP Seals

Resin/Fabric guide band damaged by severe metallic contamination

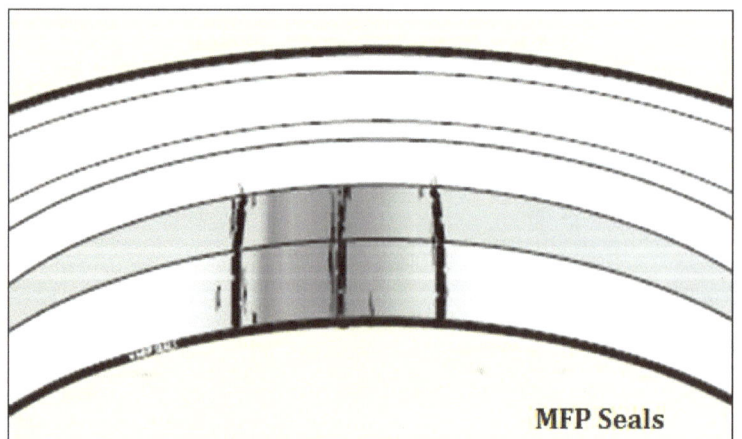

Dynamic seal lip shows axial cuts and grooves

Fig. A.164 - Seal Failure due to Contamination

11.13- Operational Defects - Fluid Incompatibility

Failure Source: incompatibility with the operating hydraulic fluid.

Failure Mode: As shown in Fig. A.165, a hydraulic seal lost its flexibility and cracks are formed.

Suggested Solution: Make sure hydraulic seals and hydraulic fluids are compatible.

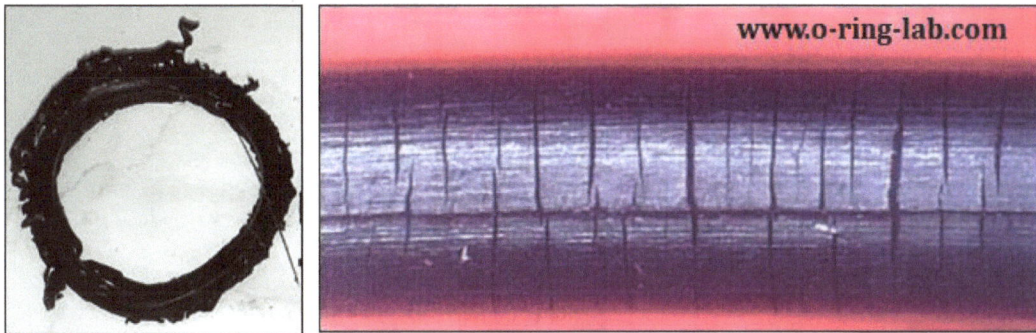

Fig. A.165 - Seal Failure due to Hydraulic Fluid Incompatibility (www.o-ring-lab.com)

11.14- Operational Defects - Chemical Attack

Failure Source: Chemical interaction between the seal and the hydraulic fluid

Failure Mode: As shown in Fig. A.166, a hydraulic seal was subjected to unfavorable defects such as excessive hardening, softening, swelling, and shrinkage.

Suggested Solution:
- Make sure acidity of the hydraulic fluid is within allowable limits.
- Check the resistivity level of the hydraulic seals to chemical attacks.

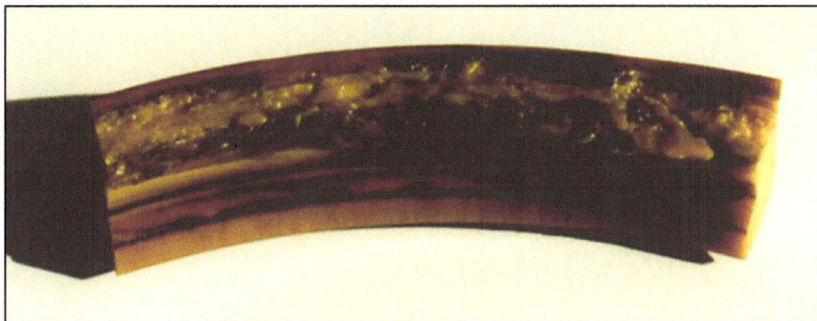

Fig. A.166 - Seal Failure due to Hydraulic Fluid Chemical Attack

11.15- Operational Defects - Hydrolysis

Failure Source: exposure to water or water-based fluids at elevated temperatures.

Failure Mode: Figures A.167 and A.168, show break-down of the seal material, loss of physical properties, cracking, and crumbling of the material due to *Hydrolysis* Failure.

Suggested Solution: Select proper seal material for water-based fluids.

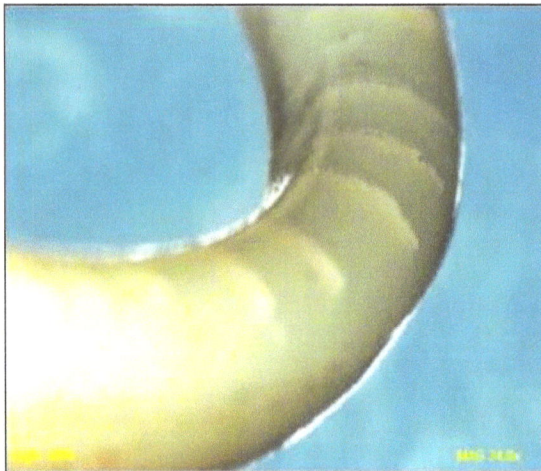

Polyurethane O-Ring

Fluorocarbon (FPM) seal

Fig. A.167 - Seals Showing Early Signs of Hydrolysis (Courtesy of System Seals Inc.)

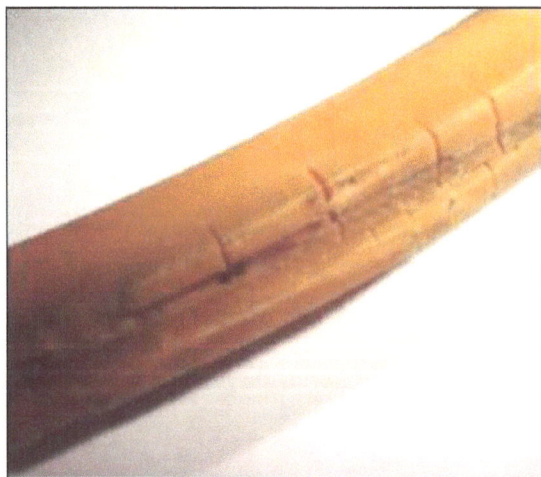

Thermoplastic elastomer seal

Fluorocarbon (FPM) seal

Fig. A.168 - Seals Showing Late Signs of Hydrolysis (Courtesy of System Seals Inc.)

11.16- Operational Defects – Explosive Decompression

Failure Source: Sudden decompression of air bubbles. Mineral oils contain, at atmospheric pressure, up to 10% by volume molecularly dissolved air. In a "saturated" condition, the dissolved air has no effect on the hydraulic oil performance. If the pressure on the oil falls, the high volume of air molecules can no longer remain in solution, and the air will separate and form bubbles. When the operating fluid is aerated, air bubbles expand in an explosive manner very quickly upon sudden pressure drops resulting in damage to the sealing elements.

Failure Mode: As shown in Figures A.169 through A.171, hydraulic seals failed due to air bubbles sudden decompression.

Standard Test Method (Explosive Decompression Test): A high pressure test rig is used to pressurize and depressurize the sealing element at a certain frequency under specified temperature.

Suggested Solution:
- Apply the required design strategies to prevent hydraulic fluid aeration.
- Apply design strategies to control the decompression rate of pressurized fluid.
- Use *Anti-Explosive Decompression* (AED) seals.

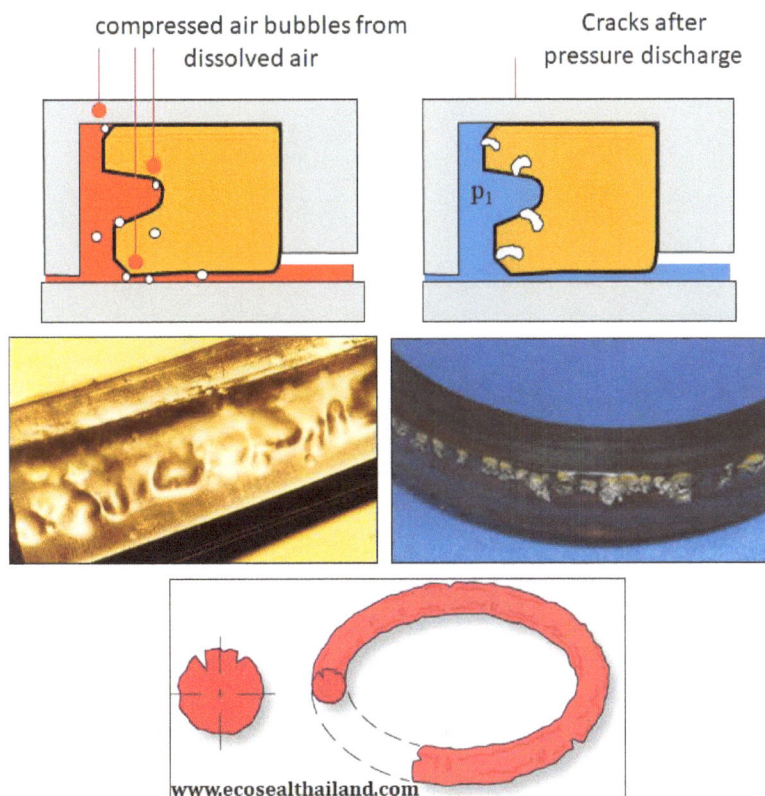

Fig. A.169 - Seal Failure due to Explosive Decompression, Example 1

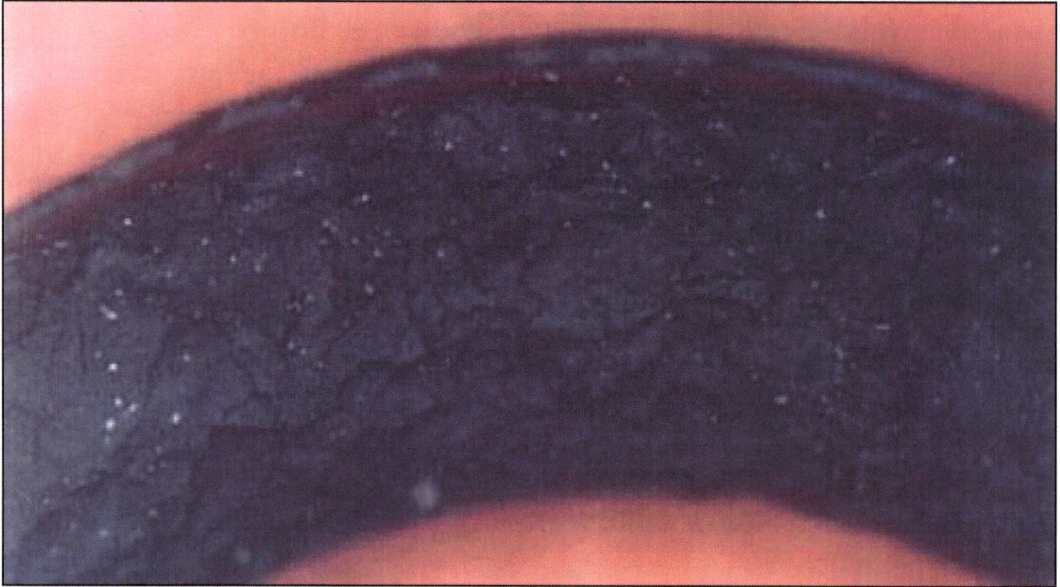

**Fig. A.170 - Seal Failure due to Explosive Decompression, Example 2
(www.o-ring-lab.com),**

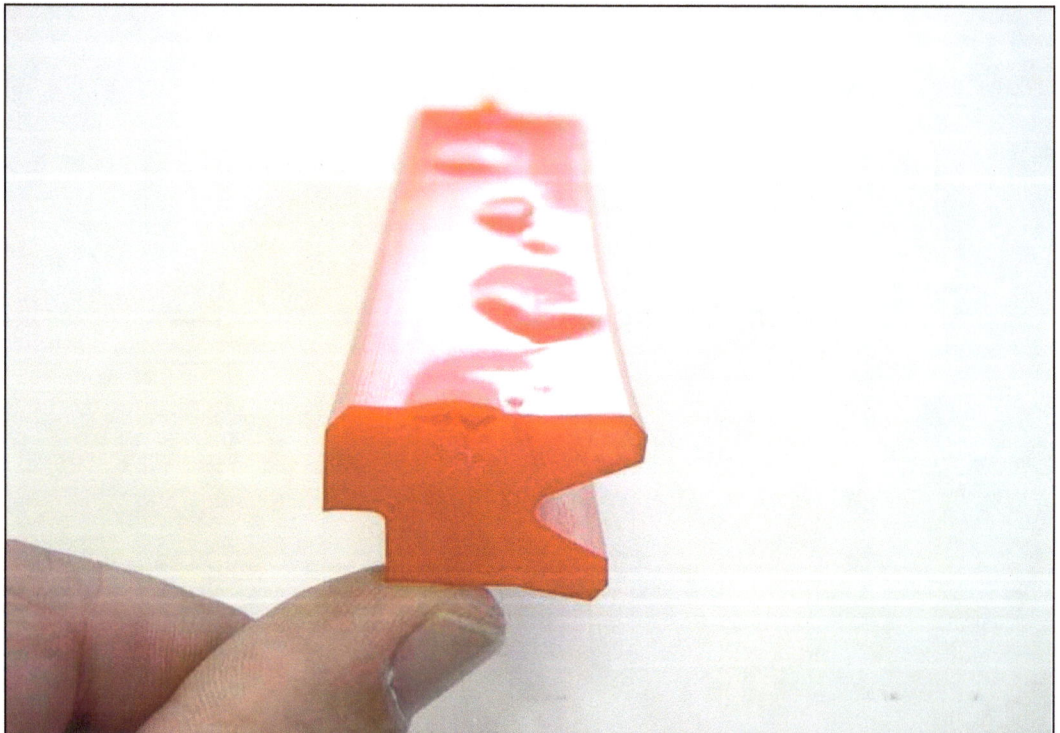

Polyurethane u-cup showing severe dieseling damage

**Fig. A.171 - Seal Failure due to Explosive Decompression, Example 3
(Courtesy of System Seals Inc.)**

11.17- Operational Defects - Dieseling

Failure Source: Sudden compression of air bubbles. When the operating fluid is aerated, air bubbles are compressed and self-ignite very quickly upon sudden pressure increase with the presence of extreme working temperature. This is known as the *Diesel Effect* like fuel burning in diesel engines.

Failure Mode: As shown in Figures A.172 and A.173, hydraulic seals are burned and damaged as a result of dieseling effect.

Suggested Solution:
- Apply the required design strategies to prevent hydraulic fluid aeration.
- Apply design strategies to control the working temperature and pressure shocks.

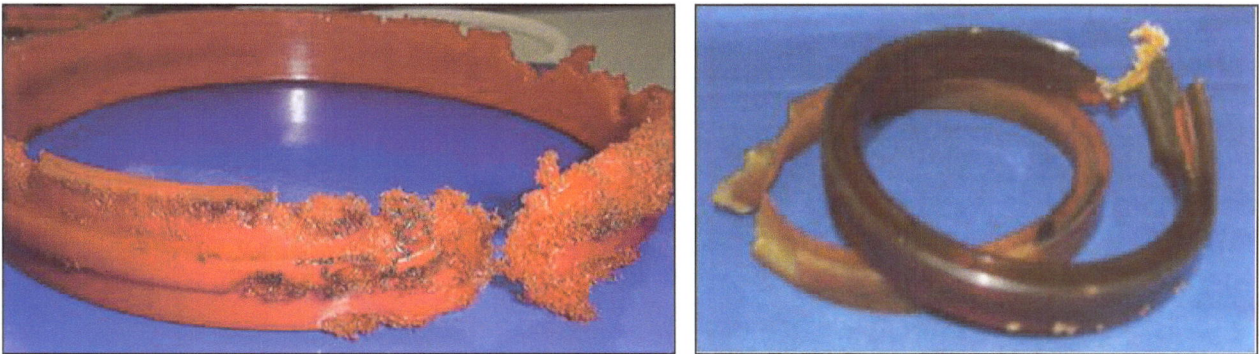

Fig. A.172 - Seal Damage due to Diesel Effect, Example 1

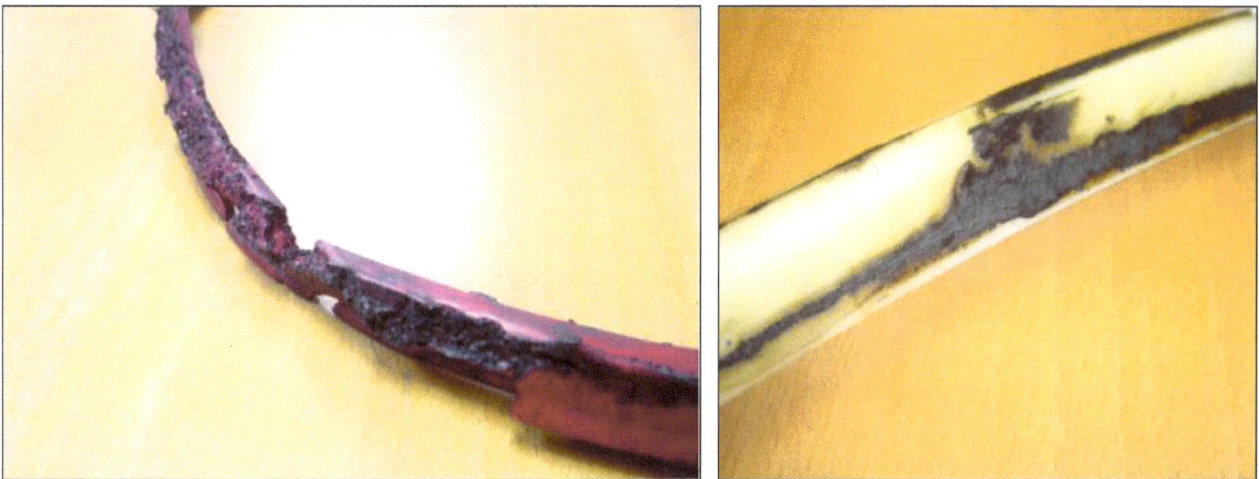

Polyurethane u-cup showing severe dieseling damage   *Nylon Back-up ring with dieseling damage*

Fig. A.173 - Seal Damage due to Diesel Effect, Example 2 (Courtesy of System Seals Inc.)

11.18- Operational Defects - Side Loading

Failure Source:
- Side loads on a cylinder piston or cylinder rod.
- Insufficient load guidance

Failure Mode: Figures A.174 shows damage to the sealing elements and the cylinder components as a result of side loading. The following are additional consequences of side loading:
- Increased gland clearance on one side only.
- Gap extrusion.
- Increased leakage.
- Uneven friction on the seal
- Rod or barrel will be galled or scored.

Suggested Solution:
- Follow the best practices for mounting the hydraulic cylinder and the attached load.
- Whenever possible, use *Stop Tubes* to reduce the effect of side loads.
- Proper seal package design that includes Guide-Rings.

Left: Severe side loading with catastrophic damage from metal-to-metal contact
Right: Side loading that wore the chrome off the rod

Fig. A.174 - Seal Damage due to Side Loading, Example 1 (Courtesy of System Seals Inc.)

11.19- Operational Defects - Vibration

Failure Source: small frequent motions which are usually encountered when equipment is in transit. Such defects are reported in hydraulic cylinders more than any other components.

Failure Mode: Excessive wear of hydraulic seals.

Suggested Solution: Apply the required design strategies to isolate the vibrations from hydraulic components.

11.20- Operational Defects - Spiral Failure

Failure Source: This failure occurs when some segments of the O-ring slide while other segments simultaneously roll. The design and operational factors which contribute to spiral failure of a seal are listed below in the order of their relative importance:

1. Speed of stroke.
2. Lack of lubrication.
4. Squeeze and softness of sealing rings.
5. Too much space for movement in the groove.
6. Temperature of operation.
7. Length of stroke.
8. Surface finish of gland.
9. Type of metal surface.
10. Side loads.
11. ID to W ratio of O-ring.
12. Contamination or gummy deposits on metal surface.
15. Eccentricity of sealing ring.
16. Stretch and softness of sealing rings.
17. No use of Back-up Rings.

Failure Mode: A unique type of failure, called *Torsional or Spiral Failure*, may occur on reciprocating dynamic sealing rings of different cross section. This failure was given this name because when it occurs, as shown in Fig. A.175, the seal has spiral 45-degree angle deep cuts through the crosssection in a spiral pattern.

Suggested Solution: Resolve the previously mentioned sources of such a failure.

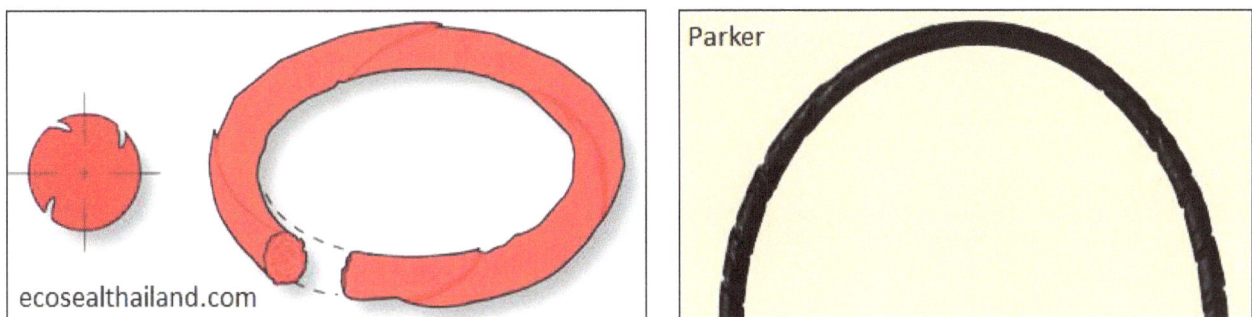

Fig. A.175 - Sealing Ring Spiral Failure (ecosealthailand.com)

11.21- Operational Defects - Seal Wear

Failure Source: The wear pattern should be even and consistent around the circumference of the dynamic lip. A small amount of even wear will not drastically affect seal performance; however, if the wear patterns are uneven or grooved, or if the amount of wear is excessive, performance may be dramatically reduced. Table A.13 lists the factors that influence seal wear.

Factors that Influence Seal Wear	
Rough surface finish	Excessive abrasion may occur above 12 µin Ra
Ultra-smooth surface finish	Surface finishes below 2 µm Ra can create aggressive seal wear due to lack of lubrication
High pressure	Increases the radial force of the seal against the dynamic surface
High temperature	While hot, materials soften, thus reducing tensile strength
Poor fluid lubricity	Increases friction and temperature at sealing contact point
Tensile strength of seal compound	Higher tensile strength increases the material's resistance to tearing and abrading
Fluid incompatibility	Softening of seal compound leads to reduced tensile strength
Coefficient of friction of seal compound	Higher coefficient materials generate higher frictional forces
Abrasive contamination	Creates grooves in the lip, scores the sealing surface and forms leak paths
Extremely hard sealing surface	Sharp peaks on hard surfaces will not be rounded off during normal contact with the wear rings and seals, accelerating wear conditions

Table A.13 - Factors Affecting Seal Wear (Courtesy of Parker)

Failure Mode: As shown in Fig. A.176, uneven, grooved, or excessive seal wear in dynamic application.

Suggested Solution: Resolve the previously mentioned sources of such a failure.

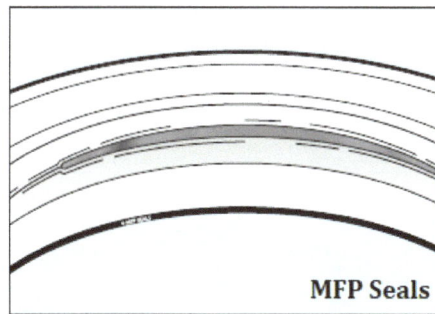

Only one side of the dynamic lip is showing excessive wear.

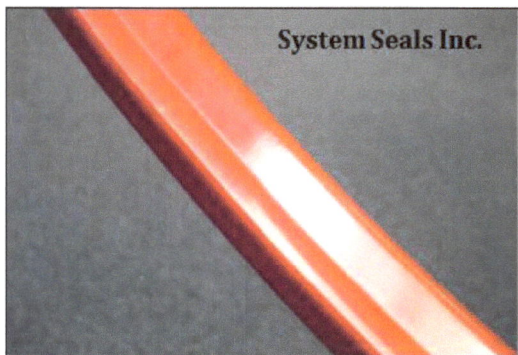

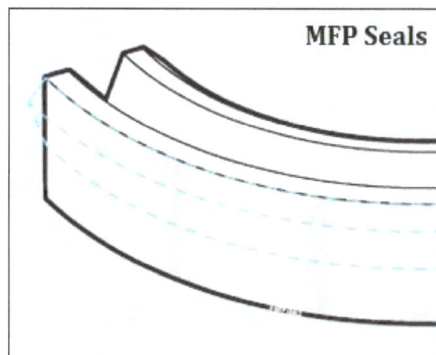

Left: Polyurethane Rod Seal with Shiny and Smooth Surface from a Dry Running Condition
Right: The dynamic face of the seal is worn to a glossy mirror like shine.

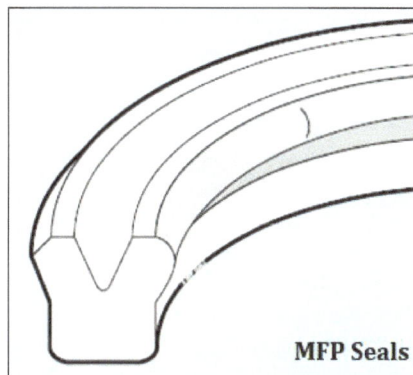

The dynamic lip is worn to a rounded, egg-shape.

Fig. A.176 - Failure Modes of Hydraulic Seals due to Wear (Courtesy of MFP Seals)

11.22- Operational Defects - Fatigue

Failure Source: Cold startup and/or exposure to cyclic pressure with high frequency.

Failure Mode: Figure A.177 shows that the V portion of the seal shows long cracks or splits.

Suggested Solution: proper selection of seal design and material. Control the startup temperature.

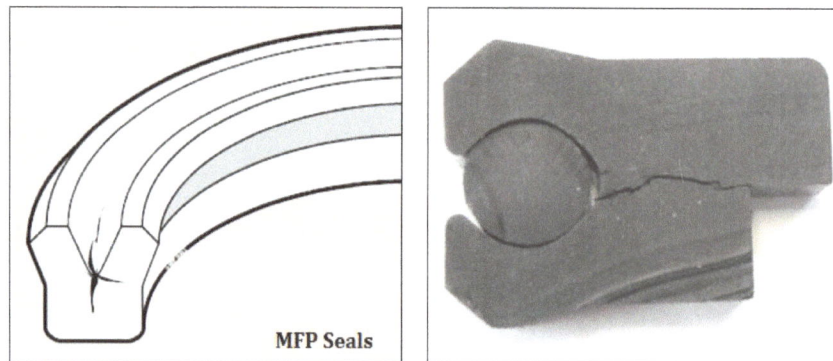

Fig. A.177 - Profile Section of an O-Ring Loaded U-Cup with Flex Fatigue Cracking (Courtesy of System Seals Inc.)

11.23- Normal Aging Defects - Hardening

Failure Source: Normal *Aging*.

Failure Mode: Figure A.178 shows cracks on the inner circumference of the sealing ring.

Suggested Solution: Use the seals within their estimated life time.

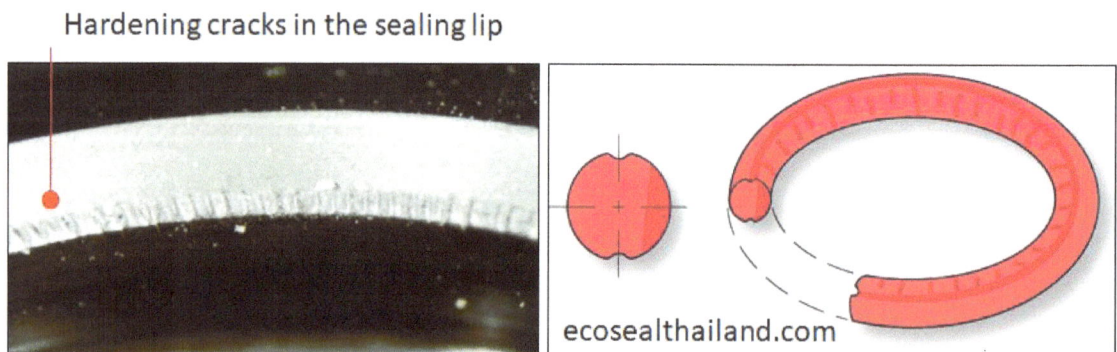

Fig. A.178 - Sealing Ring Hardening due to Normal Aging

11.24- Normal Aging Defects - Splits

Failure Source: Expired *Shelf Life*.

Failure Mode: Figure A.179 material splits through the seal body.

Suggested Solution: Use the seals within their estimated life time.

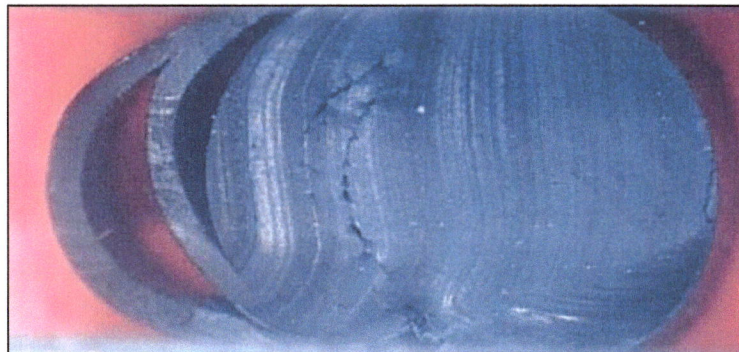

Fig. A.179 - Failure due to Normal Aging (www.o-ring-lab.com)

11.25- Storage Defects - Swelling

Failure Source: Elastomers have a higher coefficient of thermal expansion than steel. This means the seal will expand more when it is hot. Same problem could happen due to fluid incompatibility or chemical attack.

Failure Mode: Figure A.180 shows a seal absorbed the surrounding water like a sponge and swells to the point of malfunction. Figure A.181 shows a case of wiper swelling.

Suggested Solution:
- Adjust the humidity level at the storage space.
- Test the volume change of hydraulic seals periodically.

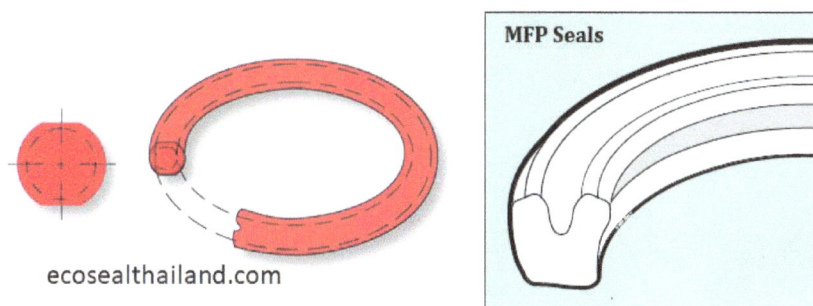

Fig. A.180 - Sealing Ring Excessively Swells due to Humid Storage Space (ecosealthailand.com)

Fig. A.181 - Swelling Failure of a wiper (Courtesy of System Seals Inc.)

11.26- Storage Defects - Ozone Cracking

Failure Source: This damage is a result of exposure of a sealing ring to Ozone for several weeks without protection.

Failure Mode: Figure A.182 shows many small surface cracks perpendicular to the direction of stress.

Suggested Solution: Provide proper protection against Ozone.

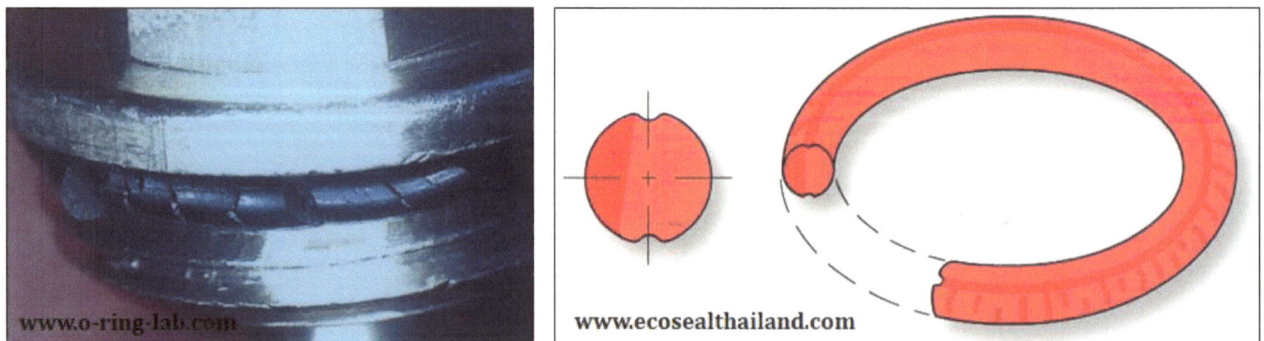

Fig. A.182 - Sealing Ring Surface Cracking due to Exposure to Ozone (ecosealthailand.com)

APPENDIX A: LIST OF FIGURES

Chapter 01: Introduction to Hydraulic Sealing Elements

Chapter 02: Sealing Rings

Chapter 03: Hydraulic Seals

Chapter 07: Sealing Solutions for Hydraulic Cylinders

Chapter 08: Sealing Solutions for Rotational Shafts

Chapter 09: Best Practices for Hydraulic Seals Installation

Chapter 10: Best Practices for Hydraulic Seals Storage

Chapter 11: Best Practices for Hydraulic Seals Failure Analysis

APPENDIX B: LIST OF TABLES

APPENDIX C: LIST OF STANDARD TEST METHODS

INDEX

X

Y

www.ingramcontent.com/pod-product-compliance
Lightning Source LLC
Chambersburg PA
CBHW051600190326
41458CB00029B/6494